FIRST AID IN SCIENCE

Robert Sulley

 HODDER EDUCATION
AN HACHETTE UK COMPANY

Acknowledgements

With thanks to Sally for her helpful comments on the manuscript, and to Eleanor, Gina and Laurice for their editorial input and advice. Thanks also to Fiona, Rosa and David for their support while I was writing this book.

Hachette UK's policy is to use papers that are natural, renewable and recyclable products and made from wood grown in sustainable forests. The logging and manufacturing processes are expected to conform to the environmental regulations of the country of origin.

Orders: please contact Hachette UK Distribution, Hely Hutchinson Centre, Milton Road, Didcot, Oxfordshire, OX11 7HH. Telephone: +44 (0)1235 827827. Email education@hachette.co.uk. Lines are open from 9 a.m. to 5 p.m., Monday to Friday. You can also order through our website: www.hoddereducation.com

© Robert Sulley 2012
First published in 2012 by
Hodder Education,
An Hachette UK Company
338 Euston Road
London NW1 3BH

Impression number   8
Year   2023

Illustrations by Pantek Media, Barking Dog Art and Robert Hichens Designs
Typeset in ITC Garamond BT 11/14pt by Pantek Media, Maidstone, Kent.
Printed and bound by CPI Group (UK) Ltd, Croydon, CR0 4YY

A catalogue record for this title is available from the British Library

ISBN: 978 1444 168914

Contents

Page

Living Things

1 Life and Living Things

Life Processes

Have you ever thought about the things we need to stay alive? We all need to eat food and breathe in oxygen from the air. Humans are animals, and all animals need to eat food and use oxygen. The scientific word for eating food is **nutrition**. The scientific word for using oxygen is **respiration**.

Nutrition and respiration are **life processes**. There are seven life processes. Animals and plants do all of them. Something is only alive if it does all seven life processes.

Here is a table with all seven life processes.

Life process	What happens?	How do animals do it?	How do plants do it?
movement	Living things move their bodies.	Animals move around to find food and to get from place to place.	Plants do not move around like animals, but they can move some of their parts. Leaves and flowers turn towards light and some flowers close up at night.
respiration	Respiration is using oxygen to turn food into energy.	Many animals get oxygen by breathing air into their lungs, or passing water over their gills.	Plants get oxygen through their leaves.
growth	Living things get bigger, for at least part of their lives.	Animals grow from babies to adults.	Plants grow from small seedlings into bigger plants.

Life process	What happens?	How do animals do it?	How do plants do it?
reproduction	Reproduction is when a living thing makes a new living thing like itself.	Animals have babies.	Plants have seeds. New plants grow from the seeds.
excretion	Excretion means getting rid of waste.	All animals get rid of waste gases, waste chemicals and surplus water. For example, humans go to the toilet to get rid of waste in urine.	Plants get rid of waste gases through their leaves.
nutrition	Nutrition is getting food.	Animals get their food by eating plants or other animals.	Plants make their own food using sunlight.
sensitivity	Sensitivity is feeling, or sensing, things.	Animals feel things such as heat.	Plants sense light and turn towards it.

One way to remember the seven life processes is to remember an imaginary teacher called Mr Grens. Each letter in Mr Grens' name will remind you of one of the seven life processes:

M – Movement

R – Respiration

G – Growth

R – Reproduction

E – Excretion

N – Nutrition

S – Sensitivity

All living things are made from **cells**. Cells are very small. Your body is made from millions of cells. Cells are too small to see unless we have a microscope.

Exercise 1

Here is a paragraph about life processes. Fill in the gaps using these words:

reproduction   nutrition   respiration   oxygen   seven

There are _____ life processes. The life process which means using food to stay alive is called _____. To turn food into energy, most plants and animals need to use _____. This life process is called _____. All living things make new living things like themselves. This life process is called _____.

By thinking about the seven life processes, we can decide whether something is living or not.

Exercise 2

Decide which of these are living and which are not living.

frog

goat

seaweed

eraser

grass

pencil

glass of water

worm

fern

car

fire

fish

rock

mosquito

lizard

tree

chopped wood

3

Classifying Living Things

There are millions of different kinds of living things, so scientists put them into groups. We put living things that are like each other into the same group. The five main groups of living things are:

| Plants | Animals | Fungi | Protoctista | Monera |

These five groups are known as **kingdoms**. Protoctista and Monera are tiny living things – many are made up of just one cell. They are often called **microbes** or **micro-organisms**. The microbe kingdoms include bacteria and the blood parasites that cause malaria. The three kingdoms that we see most often are plants, animals and fungi.

Exercise 3

Complete the table shown below by writing the name of each of these living things in the correct column.

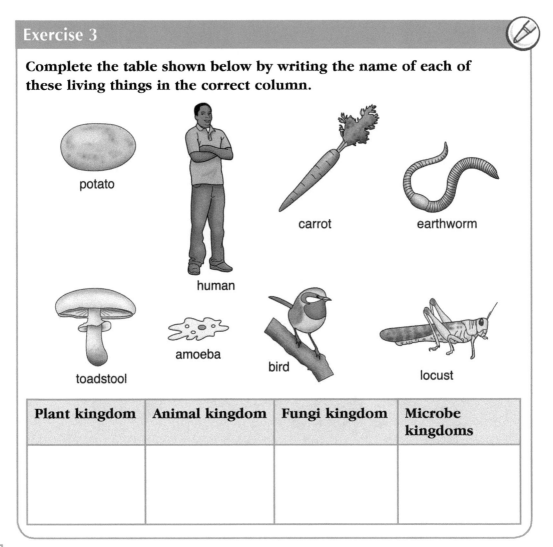

Plant kingdom	Animal kingdom	Fungi kingdom	Microbe kingdoms

The Animal Kingdom

The animal kingdom can be split into two groups. All the animals in the world can be put into one of the two groups. The two groups are:

◆ animals with a backbone – these are called **vertebrates**
◆ animals without a backbone – these are called **invertebrates.**

Vertebrates

Vertebrates all have backbones. They are divided into five smaller groups.
All the animals in each group are similar. Here are the five groups of vertebrates.

Bony fish	Amphibians	Reptiles	Birds	Mammals
Scales, fins Eggs laid in water	Smooth skin Eggs laid in water	Scales Soft-shelled eggs laid on land	Feathers Hard-shelled eggs	Hair Suckle young with milk

Invertebrates

Invertebrates don't have backbones. They are divided into smaller groups.
All the animals in each group are similar. Here are three groups of common invertebrates.

Insects	Arachnids	Molluscs
Six legs, three body parts	Eight legs, two body parts	Soft body, often with a shell

Exercise 4

Are these animals vertebrates or invertebrates? If you are unsure, think about whether they have a backbone and which other animals they are like.

a) horse   b) frog
c) worm   d) eagle
e) ant   f) bee
g) human

The Plant Kingdom

We can divide all the plants in the world into two groups by asking whether the plant has flowers. Lots of plants have flowers, some of which might surprise you. For example, many trees have flowers. If we ask this question we can then put every plant into one of two groups:

◆ **flowering plants** (plants with flowers)
◆ **non-flowering plants** (plants without flowers).

The non-flowering plants have four main groups called algae, mosses, ferns and conifers.

Here are some examples of flowering plants and non-flowering plants.

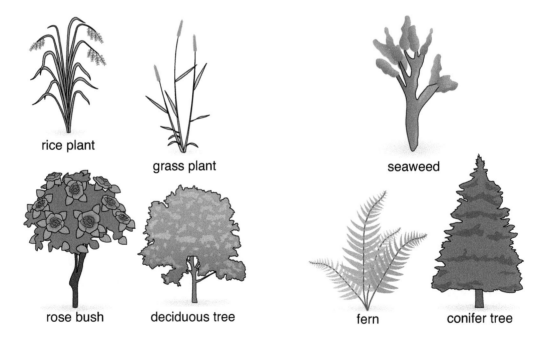

rice plant

grass plant

seaweed

rose bush   deciduous tree

fern   conifer tree

flowering plants   **non-flowering plants**

Remember!

- There are seven life processes: movement, respiration, growth, reproduction, excretion, nutrition, sensitivity.
- Something is only alive if it does all seven life processes.
- All living things are made from cells. Cells are very small. We need a microscope to see them.
- Living things can be grouped into five main groups called kingdoms: plants, animals, fungi, Protoctista and Monera. The last two are made up of micro-organisms, which are often called microbes, such as bacteria.
- The animal kingdom can be divided into two main groups – vertebrates (animals with a backbone) and invertebrates (animals without a backbone).
- The plant kingdom can be divided into two main groups – flowering plants and non-flowering plants.

Revision Test   on Life and Living Things

Now that you have worked your way through the chapter, try this revision test. The answers are in the answer book.

1. Name the seven life processes.
2. Which life process means using oxygen to get energy from food?
3. Which life process makes new living things?
4. Which life process means using food to stay alive?
5. Which life process means that living things get bigger?
6. To which kingdom of living things does a hummingbird belong?
7. To which kingdom of living things does seaweed belong?
8. Which of these animals are vertebrates (have a backbone) and which are invertebrates (don't have a backbone)?

   moth   worm   dog   spider   shark

9. Mammals are a group of vertebrates with lungs that breathe in oxygen, and give birth to live babies (not eggs). Which of these animals are mammals?

   tortoise   human   monkey   parrot   dolphin

10. Frances finds an invertebrate on her way to school. It has eight legs and two body parts. What kind of invertebrate is it – an insect, an arachnid or a mollusc?

2 Plants

Parts of the Plant

Here is a drawing showing the two main parts of a plant.

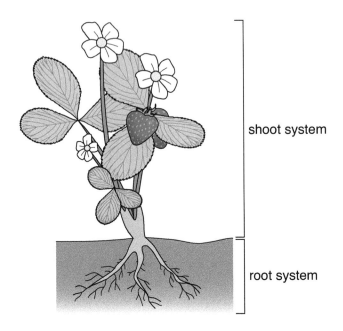

shoot system

root system

The part of the plant that grows below the ground is the **root system**.
The **shoot system** of a plant is made up of all the parts above the ground.
The stem, the leaves, the flowers and the fruits are all part of the shoot system.
These are the parts that we can see.

We are going to look at each part more closely to see what it does.

Roots

Roots do three main jobs:

◆ **Roots take in water and minerals** from the soil. Water and minerals are very important for plants to keep them strong and healthy. Minerals are natural chemicals that provide goodness for the plant. When roots take in water and minerals like this, we say they **absorb** the water and minerals.
◆ **Roots keep the plant in place**, like a ship's anchor. Large trees have deep roots to stop them blowing over when the weather is very windy.
◆ **Roots sometimes store food** for the plant.

The three main types of roots are tap roots, fibrous roots and storage roots.

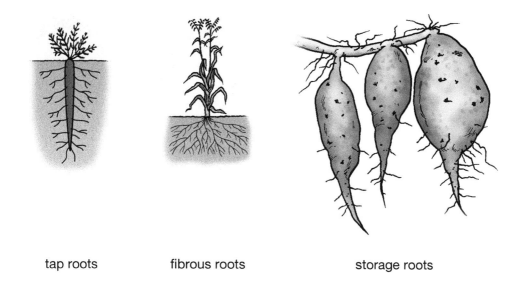

tap roots   fibrous roots   storage roots

Tap roots

A **tap root** is a strong main root that grows straight down into the soil. Smaller roots grow on the side of the main root. Tap roots hold plants strongly in place and they can absorb water and minerals from deep in the soil.

Fibrous roots

Fibrous roots are made up of many small root branches that spread through the soil. They can absorb water and minerals from a large volume of soil and from near the surface. They are good at collecting rain water from near the surface when it rains. Most grasses have fibrous roots.

Storage roots

Some plants have **storage roots** that swell to store food for the plant. Sweet potato and cassava both have roots that do this. Storage roots can be used as food by humans and other animals.

Exercise 1

a) What are the three main jobs of plant roots?
b) Name the three main types of root.
c) Give an example of a plant that has a storage root.
d) Which type of root is best for absorbing water and minerals from deep in the soil?

Stems

The **stem** of a plant is part of the shoot system. The stem grows towards the sunlight. There are different types of stem. Big plants such as trees need strong stems for support. We call these trunks. Some stems spread out, some stems climb up other plants or buildings, some stems creep along the ground. Here are some examples of stems.

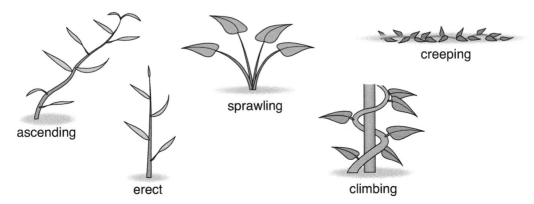

ascending

erect

sprawling

climbing

creeping

Leaves

Here is the structure of a leaf with the main parts labelled.

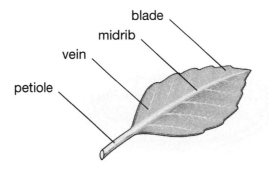

blade

midrib

vein

petiole

- The **blade** is thin and flat.
- The **petiole** is the stalk attaching the leaf to the stem.
- The **midrib** carries water to the leaf. It also carries dissolved sugar to and from the leaf.
- The **veins** branch off from the midrib, carrying water to all the cells in the leaf. The veins also carry dissolved sugar to and from leaf cells.

Leaves come in different shapes and sizes but they all do the same jobs.

Photosynthesis

Leaves make food for the plant by taking in sunlight. They do this by using the sunlight to turn water and a gas from the air called **carbon dioxide** into sugar. The plant uses the sugar for food. When a leaf uses sunlight like this it is called **photosynthesis**.

Transpiration

Leaves also lose water to the air through tiny holes in their surface. This is called **transpiration**. This makes room for more water in the plant and so acts like a pump, pulling more water up from the roots through the plant. The water carries minerals from the soil, which the plant needs to grow healthily.

Excretion

Leaves also get rid of waste gases through the tiny holes in the leaves. This is called **excretion**.

Exercise 2

Fill in the gaps in the paragraph below using the words given.

petiole   water   photosynthesis   stem   excretion   transpiration

Leaves make food for the plant using sunlight. This is called _____.
Leaves lose water through tiny holes in the surface. This draws more water through the plant. This process is called _____. Leaves also use the tiny holes to get rid of waste gases. This process is called _____. The _____ is the stalk attaching the leaf to the _____. The midrib carries _____ to the leaf, and dissolved sugar to and from the leaf.

Flowers

Flowers make the life process of reproduction possible for the plant. Flowers come in many different colours, shapes and sizes, and they have many different parts.

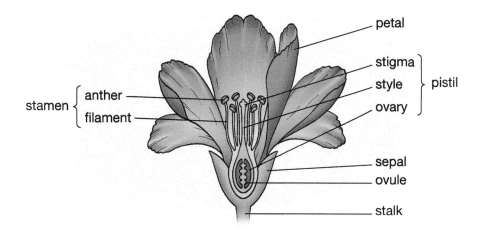

◆ The **stalk** attaches the flower to the stem of the plant, and supports the flower.

◆ The **sepals** cover the bud when the flower is growing, to protect the flower.

◆ The **petals** are often brightly coloured and have a scent to attract insects to the flower.

◆ The **stamen** is the male part of the flower. It has an **anther**, which makes pollen, and a **filament**, which holds up the anther.

◆ The **pistil** is sometimes called the **carpel**. It is the female part of the flower. It has an **ovary**, which has **ovules** inside it. The ovary grows into the fruit of the plant and the ovules grow into seeds. The pistil also has a **stigma** which is sticky to collect pollen, and a **style** which joins the stigma to the ovary.

Reproduction in Flowering Plants

We learnt in the last chapter that all living things reproduce themselves. This life process is known as reproduction. Plants can reproduce by making seeds. The seeds can then grow into a new plant. Two steps are needed for a seed to form. The two steps are **pollination** and **fertilisation**.

All flowering plants make **pollen**. Pollen is a powder made by the anther of the flower. For the plant to reproduce, the pollen has to leave the anther and land on the stigma. This is pollination. The pollen then goes down the stigma to the ovary where it joins with the ovules. The pollen grains are male sex cells and the ovules are female sex cells. When they join a seed is made. This is fertilisation.

So:

> **Pollination** is getting the pollen from the anther to the stigma.
>
> **Fertilisation** is joining the pollen and the ovule together.

Pollination

Lots of plants cannot be pollinated by their own pollen. They need pollen from flowers from a different plant of the same sort. There are two ways that pollen is carried from one plant to another.

Pollination by insects, birds or other animals

Some flowers are pollinated by animals, usually insects such as bees, and sometimes by small birds. Often they are attracted to the flowers by the brightly coloured petals. The flowers also give off a scent and make a sweet liquid called nectar, which bees like as it helps them to make honey. The bees get covered in pollen and then carry the pollen to a flower on another plant. The pollen gets stuck on the stigma of the new flower. Small birds sometimes pollinate flowers in the same way.

Pollination by the wind

Some flowers are pollinated by the wind. The flowers have long stamens that get blown by the wind. The pollen is blown from the anthers onto other plants, where it gets stuck on the stigmas of the flowers.

Exercise 3

a) Which part of a flower is often brightly coloured to help attract insects?
b) Which part of a flower is sticky to collect pollen?
c) What are the two steps needed for a seed to form?
d) Which comes first, pollination or fertilisation?
e) Name two ways that a plant can be pollinated.
f) Which part of the plant is fertilised by the pollen?

Seeds and Fruits

When the ovules have been fertilised by the pollen, they grow into **seeds**. As the seeds grow, the ovary grows around them to become the **fruit** of the plant. Fruits come in lots of different shapes and sizes. Some fruits are soft and juicy, some are hard and dry. Some fruits have only one seed inside them, like mangos, plums and avocados. Some fruits have lots of seeds inside them, such as bananas and oranges.

Dispersal

The fruits and seeds are carried away from the parent plant before they start to grow, to stop overcrowding. If the seeds grow too close to the parent plant they have to compete with the parent plant for light, water and food. So it is better for the plant if the seed is carried away. This is called **dispersal**.

Here are four ways that seed dispersal can happen.

◆ **Wind**: Some seeds are very light and are blown away from the parent plant by the wind. Others have wings, which help them to be blown away.

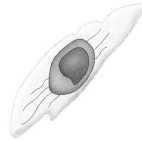
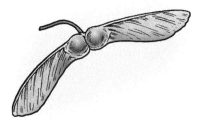

| *Tecoma* seed | sycamore seeds | wild cotton seeds |

◆ **Water**: Some fruits from waterside plants can float away to new ground. An example of a plant that disperses its seeds like this is the coconut.

◆ **Animals**: Many fruits and nuts are eaten by animals. The seeds pass right through the animal's digestive system, and leave the body in its waste. The seeds then fall onto new ground away from the parent plant. Some seeds are sticky or have hooks so they get stuck in animal fur. The animal then carries the seed to new ground before it drops off the fur.

Bananas and apples are good to eat. The seeds are inside the fruits.

Boerhavia fruits have sticky hairs. They get carried away on an animal's fur or feathers.

◆ **Explosion**: In some plants, such as peas and poppies, the fruit dries out and suddenly splits open, throwing the seeds away from the parent plant. This is called dispersal by explosion.

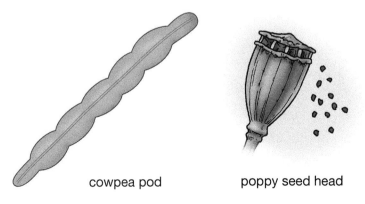

cowpea pod   poppy seed head

Germination

Once the seed has been carried away from the parent plant it can start to grow into a new plant. This is called **germination**. Each seed has a store of food inside it to help it to germinate.

Exercise 4

a) After an ovule has been fertilised, what does it grow into?
b) Which part of the plant becomes the fruit?
c) Name four ways that fruits and seeds can be dispersed.

Remember!

◆ Roots take in water and minerals from the soil, they keep the plant in place, and in some plants they store food for the plant.
◆ The three main types of roots are tap roots, fibrous roots and storage roots.
◆ Stems are part of the shoot system of a plant. Stems grow towards the sunlight.
◆ Leaves make food for the plant by using sunlight. This is called photosynthesis.
◆ Leaves lose water to the air, which pulls more water from the roots up through the plant. This is called transpiration.
◆ Leaves get rid of waste gases. This is called excretion.
◆ The main parts of a flower are the stalk, the sepals, the petals, the pistil and the stamen. You should learn these by looking at the drawing on page 11.
◆ Flowers allow plants to reproduce by making seeds. Two steps are needed to make seeds. These steps are called pollination and fertilisation.
◆ Pollination is getting the pollen from the anther to the stigma. Fertilisation is joining together the pollen and the ovule.
◆ Flowers can be pollinated by insects, by birds or by the wind.
◆ Seeds and fruit can be dispersed by the wind, by water, by animals or by explosion.
◆ Germination is when the seed starts to grow into a new plant.

Revision Test   on Plants

Now that you have worked your way through the chapter, try this revision test. The answers are in the answer book.

1. Name two things that roots take in from the soil.
2. Which type of root is best at collecting water from near the surface when it rains?

   tap root   fibrous root   storage root

3. Which part of the plant makes food for the plant? What is the name of this process?
4. Fill in the missing labels on this drawing of a leaf. Use these words:

   midrib   petiole   blade   vein

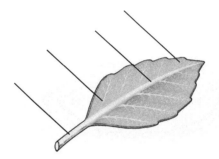

5. Fill in the missing labels on this drawing of a flower. Use these words:

   sepal   ovary   petal   stigma   anther

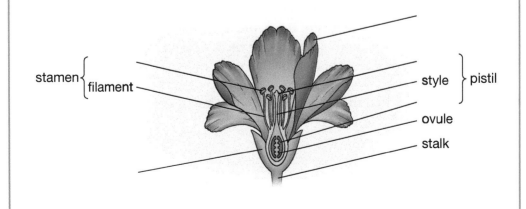

Revision Test   on Plants (continued)

6. Which part of a flower makes pollen?

   petal   anther   ovary   stigma

7. Which part of a flower is sticky to collect pollen?

   petal   anther   ovary   stigma

8. Name two ways that a plant can be pollinated.

9. Write out this sentence, and fill in the missing word:

   Fertilisation is when the pollen joins together with the _____,
   which then grows into a seed.

10. After fertilisation the seeds grow inside the ovary. What does the ovary
    then become?

11. Name two ways that seeds are dispersed by plants.

12. Each seed has a store of food inside it which helps it to grow into a new
    plant. When a seed starts to grow like this, it is called:

    respiration   germination   reproduction   transpiration

3　The Human Body

What are the Human Body Systems?

Our bodies have **systems** that help keep us alive. These body systems help us with the seven life processes that we learnt about in Chapter 1. Can you remember the seven life processes? Remember that our imaginary teacher Mr Grens will help us, as each letter of his name is the first letter of a life process: Movement, Respiration, Growth, Reproduction, Excretion, Nutrition, Sensitivity.

This table gives the seven life processes and the human body systems that help make them work. You don't need to remember all of these at this stage. In this chapter we will look at some of the most important ones that you do need to remember.

Life process	Human body system
movement	muscle system, skeletal system
respiration	respiratory system, circulatory system
growth	endocrine system
reproduction	reproductive system, endocrine system
excretion	excretory system, circulatory system
nutrition	digestive system, circulatory system
sensitivity	nervous system

The parts of the body that help these systems to work are called **organs.** For example, the **lung** is an important organ in the respiratory system that helps us get oxygen for respiration. The **stomach** is an important organ in the digestive system that helps us with nutrition.

This diagram gives the main human body systems and the organs that help them to work.

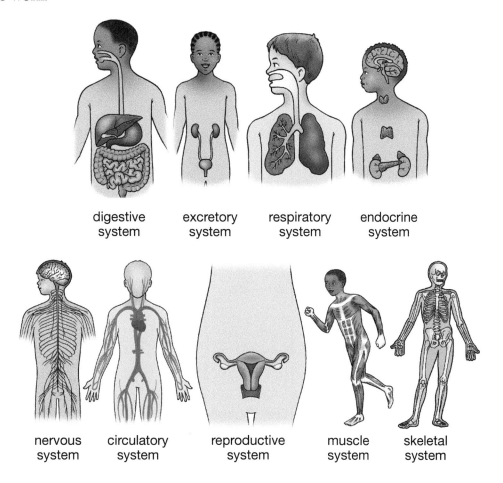

| digestive system | excretory system | respiratory system | endocrine system |

| nervous system | circulatory system | reproductive system | muscle system | skeletal system |

We are now going to look at some of the systems in more detail. We are going to look at the systems that help us with movement, respiration, nutrition, excretion and reproduction. In Chapter 7 on light and sound we will look more closely at the eye and the ear, which are parts of the nervous system.

The Skeletal System and the Muscle System

You have a **skeleton** inside your body.

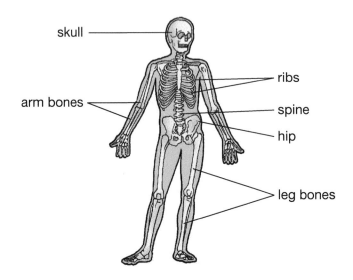

skull

ribs

arm bones

spine

hip

leg bones

Your skeleton does three important jobs:

◆ **It protects your body parts**. For example, your skull protects your brain, your ribs protect your heart and lungs, and your spine protects your spinal nerve. The spine is often called the 'backbone'.

◆ **It supports your body**. Your skeleton helps you to stand and hold up your body.

◆ **It lets you move**. Muscles are joined to your bones and the bones have **joints** between them so that the skeleton can bend. For example, your knees and elbows are joints that let you bend and move your legs and arms.

Your skeleton is covered with **muscles**.

Your muscles pull on the bones to make the skeleton bend at the joints. Muscles cannot push, they can only pull, so they work in pairs to make your skeleton move.

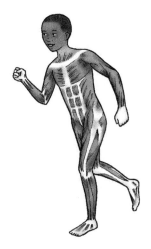

For example, there are two muscles attached to your upper arm called the biceps and the triceps. When you bend your arm they work together. To bend your arm, the biceps muscle contracts (which means it gets shorter). This pulls up the lower arm. At the same time the triceps muscle relaxes and gets longer. To straighten the arm, the triceps muscle contracts and gets shorter and the biceps muscle relaxes and gets longer.

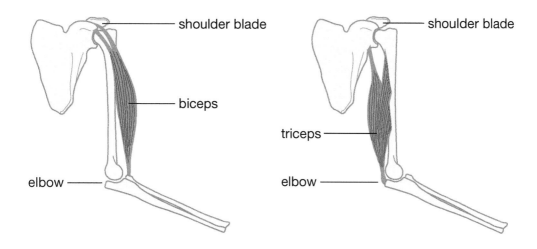

Exercise 1

a) Name three important jobs that your skeleton does.
b) Which part of your skeleton protects your heart and lungs?
c) Which part of your skeleton protects your brain?
d) How do muscles help us to move?

The Circulatory System

Your circulatory system is made up of your heart, your blood and the blood vessels that carry your blood around your body.

Your heart pumps blood around your body through your blood vessels. We say that the blood **circulates** round your body. 'Circulate' means to move around in a system, which is why this system is called the circulatory system.

Your blood carries food and oxygen around your body to your body cells, and it carries away waste from your cells.

Your blood moves through three kinds of blood vessels called arteries, veins and capillaries.

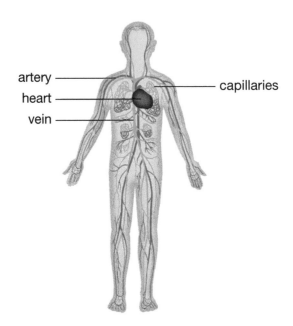

- **Arteries** carry blood from the heart to the body cells.
- **Veins** carry blood from the body cells back to the heart.
- **Capillaries** let food and gases move into and out of the blood.

Here is a diagram of your heart.

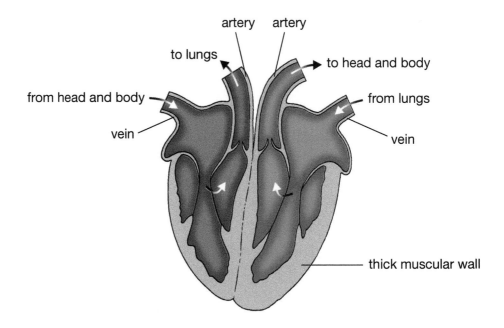

When your heart beats, it pumps blood along two arteries. One artery takes blood to the lungs. Veins bring the blood back from the lungs to the heart. The heart then pumps the blood that has come from the lungs along the other artery and around the body. Then veins bring the blood back from the body cells to the heart, and it starts again.

You can feel how fast your heart is beating by taking your **pulse**. You can feel your pulse on your wrist or on your neck. If you count how many times your pulse beats in one minute it tells you how fast your heart is beating.

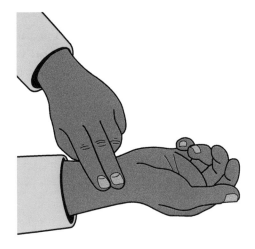

The heart beats about 70 times each minute in an adult, when the person is resting. Children's hearts beat faster. Try taking your pulse now. How fast is your heart beating?

The Respiratory System

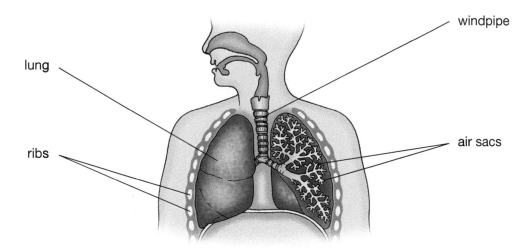

lung

windpipe

ribs

air sacs

Your lungs are like spongy bags filled with millions of tiny **air sacs**. The air sacs are covered with capillaries, which carry blood. The job of the lungs is to give oxygen from the air to your blood and to remove from your blood a waste gas called **carbon dioxide**. Oxygen moves into the blood in the capillaries and carbon dioxide moves out of the blood from the capillaries at the same time. This is called **gas exchange**. The carbon dioxide is then breathed out of your body.

When your body works hard, for example if you are running or swimming, then your muscles need more energy, so your heart beats faster and you breathe faster to get more food and oxygen to the muscles.

Exercise 2

a) Which organ pumps blood around the body?
b) Name the three types of blood vessel in your body.
c) Which type of blood vessel lets gases and food move into and out of the blood?
d) Which type of blood vessel carries blood away from the heart?
e) Which gas does the blood take from the air in your lungs?
f) Why does your heart beat faster and your breathing get faster if you start to run?

The Digestive System

The food we eat is too big to get into the cells in our bodies. It has to be broken down into simple chemicals so that it can be taken to the cells by our blood. Breaking down food in our bodies is called **digestion** and so the system that does it is called the digestive system.

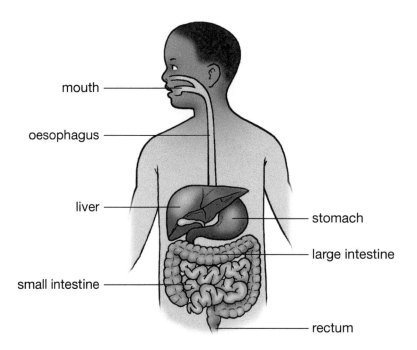

We take in the food we eat through the mouth. The food then goes down the **oesophagus** into the **stomach**. Inside the stomach there are acid juices and special chemicals called **enzymes**, which help to break down the food. The **liver** produces a liquid, called **bile**, which helps to break down fat in your food.

After the food leaves the stomach it goes into the **small intestine**. The small intestine is a tube made of muscle and it is about 8 metres long. It is wound up tightly so that it fits inside the body. Most digestion takes place in the small intestine. The muscles squeeze the food along the small intestine to mix it with more juices. Once the food is digested, it passes through the walls of the small intestine into the blood.

Any food that can't be digested, or that the body doesn't need, passes into the **large intestine** and then to the **rectum**. It then leaves the body as solid waste known as **faeces** when we go to the toilet.

The Excretory System

The body makes harmful waste as it uses energy for life processes. The body also has waste water that it doesn't need. We get rid of these using the excretory system.

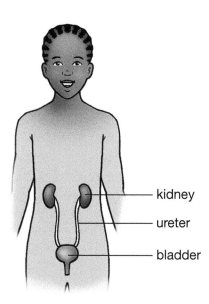

The waste produced by the body cells passes into the blood. As the blood passes through the **kidneys**, a waste called **urea** leaves the blood with some water. The mixture of urea and water is called **urine**. The urine passes from the kidney to the **bladder**. It then leaves the body as urine when we go to the toilet.

When we are hot we also excrete water and urea from the blood as it passes through the skin. Water and a little urea pass onto the skin to cool us down. We call this sweat.

When we looked at the respiratory system, we learnt that our bodies also produce a waste gas called carbon dioxide. This gas is taken out of our blood as it passes through the lungs and is then excreted from the body when we breathe out.

The Reproductive System

We learnt in Chapter 1 that reproduction is one of the seven life processes. Reproduction is when a living thing makes a new living thing like itself. Plants make seeds, which grow into new plants. Animals have babies, which grow into adult animals.

Humans have babies when an **egg** inside the mother is fertilised by a **sperm** from the father.

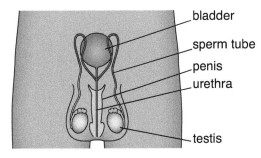
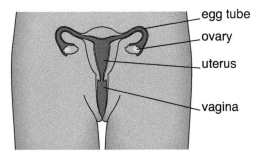

- Eggs are made in the mother's **ovaries**. Once a month an egg leaves an ovary and moves towards the **uterus**. The uterus is often called the **womb**.
- Sperm are made in the father's **testes**. (Just one is called a **testis**.) The father's testes make millions of tiny sperm each day.
- For fertilisation to happen, the sperm leave the father's penis and swim through the mother's **vagina** and uterus until they meet an egg. Fertilisation takes place when a sperm joins with the egg.
- When an egg is fertilised it grows into a tiny ball of cells called an **embryo**. The embryo grows into a baby inside the uterus.

It takes nine months for an embryo to grow into a baby ready to be born.

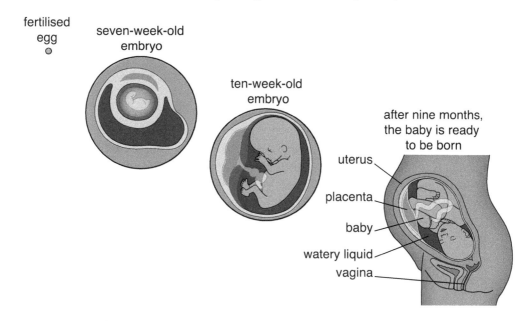

fertilised egg

seven-week-old embryo

ten-week-old embryo

after nine months, the baby is ready to be born

uterus

placenta

baby

watery liquid

vagina

As babies grow they become children and then adults. As we pass from being children to being adults we go through a stage called **puberty**. Puberty is the age when we grow quickly and our bodies change and become more like adult bodies. Boys start to make sperm and girls start to release an egg once a month from the ovaries.

Exercise 3

a) What is the name of the tube that carries food from your mouth to your stomach?
b) How does the liver help to break down food?
c) Where does most digestion take place in our bodies?
d) When food that we need has been digested, how does it get into our blood?
e) Where in the body are sperm made?
f) When a mother's egg is fertilised it grows into a tiny ball of cells. What is the tiny ball of cells called?

Remember!

- Our bodies have systems that help us with the seven life processes.
- The parts of the body that help these systems to work are called organs.
- The skeleton and muscle systems protect our bodies, support our bodies, and let us move.
- The circulatory system lets blood move around our bodies.
- The heart pumps blood through arteries, capillaries and veins.
- Arteries carry blood away from the heart. Veins carry blood back to the heart. Capillaries let food and gases move into and out of the blood.
- The lungs give oxygen from the air to our blood and remove carbon dioxide from our blood. This is called gas exchange.
- Breaking down food in our bodies so that we can use it is called digestion.
- In the stomach there are 'juices' that help break down food. The liver makes bile which helps us break down fat.
- Food passes from the stomach to the small intestine.
- Most digestion happens in the small intestine.
- Digested food passes through the wall of the small intestine into the blood.
- The kidneys help us to get rid of urea and excess water as urine.
- The reproductive system is our system for making new life.
- An egg inside the mother is fertilised by a sperm from the father to make an embryo.
- The sperm are made in the father's testes. The eggs are made in the mother's ovaries.
- It takes nine months for an embryo to grow into a baby ready to be born.

Revision Test | on The Human Body

Now that you have worked your way through the chapter, try this revision test. The answers are in the answer book.

1. Match each set of organs with the right human system. The first one is done for you.

    heart, arteries and veins —————— digestive system
    biceps and triceps   excretory system
    skull, ribs and spine   ——circulatory system
    kidneys and bladder   reproductive system
    ovaries and uterus   muscle system
    stomach and small intestine   skeletal system

2. What is the main job of the circulatory system?
3. Which type of blood vessel carries blood from the body cells back to the heart?
4. About how many times does an adult's heart beat in one minute (when the person is resting)?
5. Which gas does blood take in from the air in the lungs?
6. Which gas does blood give to the air in the lungs to be breathed out?
7. Which type of blood vessel lets gases pass between the lungs and the blood?
8. Which organ produces bile to break down fats in food?
9. Food passes through these parts of the digestive system as it is digested. Put them in the right order.

    stomach   mouth   large intestine   oesophagus   small intestine

10. Where in a human mother's body are her eggs made?
11. Where in the father's body are the sperm made?
12. How long does it normally take from fertilisation of the mother's egg before a human baby is born?

4 Nutrition and Health

To stay fit and well there are three things that we must do:

- We must eat as well as we can.
- We must stay away from things that are unhealthy.
- We must exercise as often as we can. Exercise like running or swimming helps to strengthen our muscles, including the heart, and it helps to develop our lungs.

In this chapter we are going to learn about eating well and staying healthy.

Nutrition

All living things need **nutrients**. Nutrients are the chemicals a living thing uses to grow and to repair itself. The process of feeding a living thing with nutrients is called **nutrition**. We learnt in Chapter 1 that nutrition is one of the seven life processes.

We learnt in Chapter 2 that plants use sunlight, water and carbon dioxide from the air to make the food that gives them energy. They take other nutrients from the soil through the roots. Animals, including humans, must take all of the nutrients they need by eating food. Most of our food comes from eating plants, or the bodies of other animals.

A Balanced Diet

To stay fit and well and to grow we must eat a **balanced diet**. This means we must eat all the nutrients our bodies need to stay healthy. There are five main groups of nutrients, called:

- **carbohydrates**
- **proteins**
- **fats**
- **vitamins**
- **minerals**

We must also drink **water** and eat **fibre** to stay healthy. There are two sorts of carbohydrates, called **starches** and **sugars**. We need both in our diet to stay healthy.

This diagram and table show which foods have each of these nutrients.
The table also says why we need each group of nutrients.

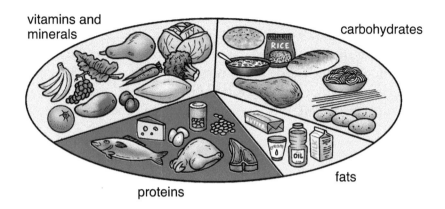

Things we need for a balanced diet	Some foods which have them	Why we need them
carbohydrates – starches	bread, rice, pasta, cereal	for energy
carbohydrates – sugars	sweets, cakes, biscuits	for energy
proteins	fish, meat, milk, eggs, beans	for cell repair and growth
fats	milk, cheese, butter, meat	for energy
vitamins and minerals	fruit, vegetables	for healthy cells, including bones, teeth and blood
fibre	fruit, vegetables, cereals, wholegrain bread	helps digestion by moving food through our intestines
water	drinks, especially water	the human body is about 70% water, so we need to drink water to stay at this level and to help keep our body temperature in balance

We can see from the table that some foods give us more than one of the nutrients we need. For example, fruit and vegetables give us vitamins, minerals and fibre. This is why it is important to eat fruit and vegetables as part of a balanced diet.

This food pyramid shows us the food groups and which nutrients they give. To have a balanced diet we must eat more of the foods at the bottom of the pyramid and less of the foods at the top. This means eating lots of bread, rice, fruit and vegetables, not quite so much fish and meat, and not very much fat or sugar. We must also drink lots of water or fruit juice and not too many sweet, fizzy drinks!

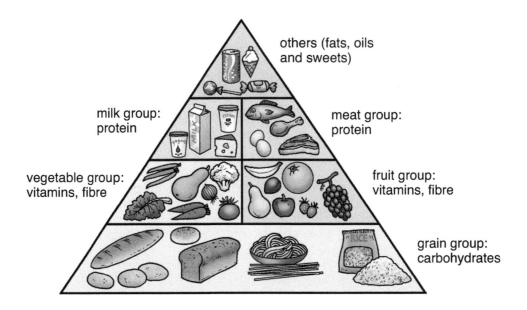

others (fats, oils and sweets)

milk group:
protein

meat group:
protein

vegetable group:
vitamins, fibre

fruit group:
vitamins, fibre

grain group:
carbohydrates

Testing for Different Nutrients

How do we know which nutrients are in a food? Scientists have found tests that will tell us. Some of the easiest tests to carry out are for carbohydrates, fats and proteins.

The next table gives simple experiments that can be done to test for carbohydrates, fats and proteins. You might not be able to try all of these at school as you may not have all the chemicals, but everyone can try the test for fat.

Nutrient	Test	What you do	Results
carbohydrates – starch	the iodine test	Add a few drops of iodine solution to the food.	If the food has starch in it, the iodine will turn from brown to black.
carbohydrates – simple sugars	Benedict's test	Mix the food with water. Add Benedict's solution and heat the mixture.	If the food has a simple sugar in it, the solution will turn from blue to orange, red, yellow or green.
fat	the grease spot test	Rub a small piece of food onto a piece of paper. Hold the paper up to the light.	A greasy mark or spot shows that the food has fat in it.
protein	the biuret test	Mix the food with water. Add a few drops of a chemical called potassium hydroxide solution, then add a few drops of copper sulphate solution and shake the mixture.	If the food has protein in it the mixture will turn purple.

Exercise 1

a) Name the five main groups of nutrients.
b) What other two things must we eat and drink to stay healthy?
c) Which two nutrient groups provide energy?
d) Why do we need to eat fibre?
e) Name two kinds of food that give us fibre.
f) What test could we do to see if a food has fat in it?
g) Which chemical do we add to food to test for starch?

Staying Healthy

If we want to stay well we must try to live a healthy life. We all get ill sometimes, but if we learn about the things that make us ill and how to avoid them we can stay healthy for longer. We are now going to look at some of the things that can make us ill.

Illness Caused by Eating Badly

If we don't eat a balanced diet we can become very ill.

Some people, especially in rich countries, make themselves ill by eating too much food and not taking enough exercise. This puts a strain on their heart and other organs. This illness is called **obesity**.

In many countries people don't have enough to eat, or cannot get a good balance of food, so they don't have a balanced diet. This makes them ill. When people become ill because they don't have a balanced diet it is called **malnutrition**. One of the most common causes of malnutrition is lack of vitamins and minerals because people don't have enough fresh fruit and vegetables to eat.

Illness Caught from Other People

We can also become ill by catching a disease from another person. These diseases are caused by microbes that are harmful to us. We call harmful microbes that cause disease **germs**. **Bacteria** and **viruses** are different kinds of germs. Germs can pass from one person to another person through the air, in water, by insect bites, or sometimes straight from the person who is ill. Diseases that pass from one person to another are called **communicable diseases**.

This table gives some examples of communicable diseases and how they are spread.

Disease	How it is spread
tuberculosis (TB)	through the air from one person to another
chicken pox	through the air from one person to another
measles	through the air from one person to another
influenza ('flu')	through the air from one person to another
typhoid	in dirty water
cholera	in dirty water
hepatitis A	in dirty water
malaria	by mosquito bites
dengue fever	by mosquito bites
HIV/AIDS	by contact with the blood of an infected person, usually through unprotected sex or by injecting with a needle that has been used by an infected person

Not all microbes are bad. Some kinds of bacteria do important jobs like helping us to make cheese and yoghurt, and helping to rot down dead organisms to put nutrients back into the soil. But to stop becoming ill we must stop the spread of harmful microbes.

Stopping the Spread of Disease

At any time there will be some people who are ill with a disease. But we can stop disease from spreading by following a few simple rules:

◆ Look after food. Cover food to stop flies from leaving germs on it. Store food in a fridge if there is one, and always keep fresh meat cool or cold. Heat food properly when cooking. Make sure rotten food is thrown away. All these things will help stop the spread of germs on food.
◆ Keep your community clean and tidy. Get rid of rubbish properly. Don't put rubbish into rivers or lakes. Only drink clean water. Control insects such as flies and mosquitoes. These things will help stop the spread of disease.
◆ Look after your **hygiene**. This means keeping yourself clean and making sure you wash your hands after going to the toilet. This will help stop the spread of germs.

- Don't sneeze or cough on other people. This will help stop the spread of diseases that are carried through the air.
- Have **vaccinations**. Vaccinations are injections given by doctors and nurses, which protect us against some diseases. We should have vaccinations and help others to have them too.

Epidemics

Sometimes a disease spreads to a lot of people at the same time. More people than usual catch the disease. When this happens, it is called an **epidemic**. Here are some of the things that can cause an epidemic:

- People stop having vaccinations.
- Water or food becomes dirty and covered in germs.
- People allow rubbish to pile up near their homes, which causes germs to spread.
- A germ changes to a new form. When this happens it means that new medicines or a new vaccination have to be made.

Almost every year a new form of the virus that causes flu develops somewhere in the world. When this happens scientists have to make new medicines and vaccines to protect old people and sick people against the new flu virus.

If an epidemic gets really big and spreads from country to country across the world, it is called a **pandemic**.

Exercise 2

a) What is the name of the disease people get if they don't have enough food to eat?
b) What is a communicable disease?
c) Name two ways that communicable diseases can be spread.
d) How is HIV/AIDS spread from person to person?
e) Name a disease that is spread through dirty water.
f) What is a vaccination?

Drugs

Many drugs are good for us and are used to make us better when we are ill. These drugs are known as medicines. We must always follow the instructions carefully because if we have too much or too little medicine it can make us ill instead of making us better.

Some drugs are very bad for us and can seriously damage our bodies. Some are **addictive,** which means that if you start taking them it is very difficult to stop.

Smoking causes heart attacks, blocked arteries, lung cancer and breathing problems. Tobacco has a chemical in it called nicotine which causes addiction.

Alcohol in small amounts is not as harmful as smoking, but it slows down the body and the brain. Drinking too much alcohol damages the liver, the heart, and the stomach. It also causes blood pressure to rise.

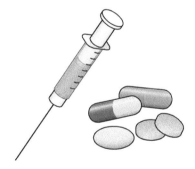

Solvents, such as glue and paint, are extremely dangerous when sniffed. They damage the brain and are very addictive.

Illegal drugs, such as cocaine and heroin, are very dangerous and can kill if too much is taken. They are very addictive and cause heart damage, lung damage and brain damage.

Exercise 3

a) Name two health problems caused by smoking.
b) What is the name of the chemical in tobacco that causes addiction?
c) Which part of the body is damaged by solvents?
d) Name two body organs that can be damaged by drinking too much alcohol.

Remember!

◆ All living things need nutrients.
◆ Feeding living things with nutrients is called nutrition. This is one of the seven life processes.
◆ To stay fit and well we must eat a balanced diet.
◆ A balanced diet has carbohydrates (for energy), proteins (for repair and growth), fats (for energy), vitamins and minerals (for healthy cells), fibre (to help digest food) and water (to keep our water level right and to help balance our temperature).
◆ Eating too much food causes obesity.
◆ Eating a diet that lacks important nutrients causes malnutrition.
◆ Diseases that pass from one person to another are called communicable diseases.
◆ Communicable diseases can pass from one person to another through the air, in dirty water or by insect bites. HIV/AIDS is passed from one person to another by contact with the blood of an infected person.
◆ We can stop the spread of disease by looking after food and cooking it properly, by keeping ourselves and our communities clean and tidy, and by having vaccinations.
◆ When more people than normal catch a disease it is called an epidemic.
◆ Drugs that are taken to treat illness are called medicines.
◆ Many drugs, such as tobacco and illegal drugs, are bad for us.

Revision Test on Nutrition and Health

Now that you have worked your way through the chapter, try this revision test. The answers are in the answers book.

1. Which one of these nutrients gives us energy?

   vitamins   proteins   carbohydrates   fibre

2. Name two types of food that give us protein.
3. Which one of these foods is not a good source of vitamins?

   oranges   rice   carrots   bananas

4. How much of the human body is made up of water?

   about 10%   about 30%   about 50%   about 70%   about 90%

5. Which of these tests is a test for proteins?

   the iodine test   the biuret test   Benedict's test   the grease spot test

6. Name two diseases that are spread from person to person through the air.
7. Name a disease that is spread by an insect bite.
8. Which one of these diseases is **not** a communicable disease?

   chicken pox   typhoid   lung cancer   HIV/AIDS   malaria

9. If a disease spreads quickly and is caught by more people than normal it is called:

   a plague   a vaccination   an epidemic   a fever

10. Smoking and drinking too much alcohol both damage organs in the body. Which one damages the liver and which one causes lung cancer?

Energy, Forces and Motion

5 | Forces

We can't see **forces**, but we can see the effects forces have on us and on the things around us. Forces are nearly always **pushes** or **pulls**, and they always work in a certain direction.

What Do Forces Do?

Forces can make objects do five things. It is useful to remember these.

Speed up
A force from the engine is making this rocket speed up.

Slow down
Air resistance and friction are two forces that make things slow down. Here, the air resistance pushing against the truck slows it down.

Turn
The pushing force from the person's hand makes the spanner turn.

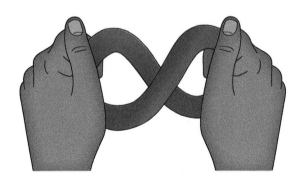

Change direction
The force from the bat makes the ball change direction.

Change shape
A force from the person's hands can make this rubber toy change shape by stretching, twisting, squashing, or bending it.

Balanced and Unbalanced Forces

Look at these groups of children pulling on a rope. In drawing A the two groups are pulling on the rope with equal force. This means that the forces acting in each direction are equal, and so the flag in the middle of the rope doesn't move. The forces acting in each direction are balanced.

In drawing B the group on the right is pulling with more force than the group on the left. The forces are no longer balanced and so the flag moves to the right, in the direction of the strongest force.

Air Resistance and Friction

Two forces that slow down moving objects are **air resistance** and **friction**. Air resistance is sometimes called 'drag'. Air resistance pushes against objects that are moving through the air. If things need to go fast they need to be shaped so that the air flows around them easily. This will decrease the air resistance. This is why racing cyclists have special equipment to reduce air resistance, and why racing cars are shaped to make air resistance less.

racing bicycles and cars are designed so that there is less air resistance

Friction is a force that slows down objects that are sliding past each other. If we push a book across a desk, friction is the force that stops the book sliding forever. If we are riding a bike, then friction is the force between the tyres and the road that slows us down when we stop pedalling. We need friction to help us grip things. Without friction our tyres would not grip the road and we wouldn't be able to move forwards. Friction is also the force between the brakes and the wheel that helps us to slow down. A rough surface gives more friction than a smooth surface, so if we want something to slide or move more easily we should make the surfaces smooth.

friction between the brakes and the wheel slows the wheel down

friction between the rock and the ground makes it hard to move the rock along

Exercise 1

a) What are the five things that forces can make objects do?

b) Fill in the gaps in the sentences below using three of these words:

force   unbalanced   weaker   balanced   stronger

Lucy and George are pulling in opposite directions on each end of a skipping rope. To start with they don't move so the forces must be _____. After a while they start to move in the direction George is pulling. The forces must now be _____. George must be pulling with a _____ force than Lucy.

c) Give an example of a force that will make an object slow down.

d) Here is a picture of a children's slide. What sort of surface does the slide need to make it easy for children to slide down it?

Forces and Turning

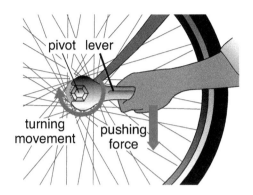

A force can make an object turn. Objects turn around a point called the **pivot**. For example, if we turn a nut around a bolt with a spanner we are using a force on the end of the spanner to turn the nut around the bolt. The bolt is the pivot. The spanner works like a **lever**.

Here is another example. If we push a door, we are using a force to make the door turn around the hinges. The hinges are the pivot.

Lots of things we use every day use a force to turn something around a pivot. The drawings below show two more examples.

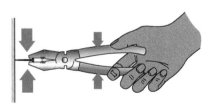
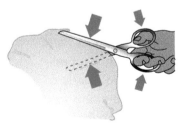

Magnetism and Gravity

The forces we have looked at so far in this chapter are all called **contact forces** because they all need objects to be touching each other for the force to work. For example, the children pulling on the rope are touching the rope. They are in contact with it. The racing bike is touching the ground to get friction and the air is touching the cyclist to give air resistance. When we turn a spanner or a door we touch them so that we can push on them.

There are other forces called **non-contact forces**. Two examples are magnetic force, known as magnetism, and gravity. These are forces that can make an object move without touching it.

Magnetism

Magnetism is the force that a magnet exerts on another magnet and on magnetic materials.

A magnet has a north pole (N) and a south pole (S).

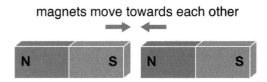

north pole   south pole

Magnetic force makes a north pole and a south pole **attract** each other. This means they are pulled towards each other.

magnets move towards each other

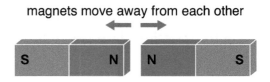

Magnetic force makes two north poles **repel** each other. This means they push each other apart.

Magnetic force makes two south poles repel each other.

magnets move away from each other

Both ends of a magnet will attract magnetic objects, such as nails, steel spoons and many other metal objects. For an object to be magnetic it must be made from iron, steel, nickel or cobalt.

magnets move away from each other

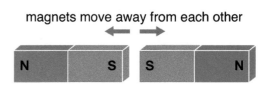

magnetic objects move
towards the magnet

Gravity

Gravity is another non-contact force. Gravity is the force that pulls objects towards the centre of the Earth. Gravity pulls objects downwards when they are in the air, in water, or on the ground. If we throw something up into the air, gravity is the force that makes it come back down again.

When we stand on the ground, gravity is pulling us towards the centre of the Earth. The surface of the Earth pushes back against our feet with an equal force so that the forces are balanced. If we rest a book on a table, gravity will pull the book down. The table pushes back against the book with an upward force so that the forces are balanced.

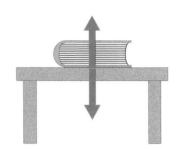

When parachutists jump out of aeroplanes, the force of gravity pulls them towards the Earth.
To stop falling too fast the parachutist opens a parachute. When the parachute opens there is a big increase in the force of air resistance, which slows down the parachutist.

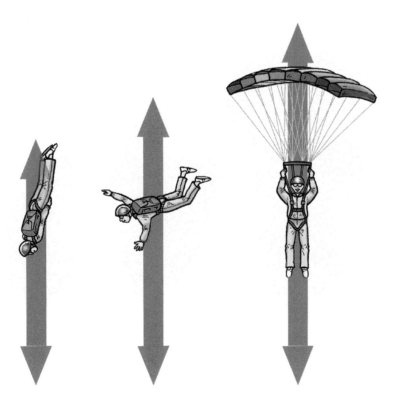

Exercise 2

a) Forces can make objects turn around a point. What name do we give to the point objects turn around?

b) Name two everyday objects that use forces to turn around a point.

c) Name two non-contact forces.

d) If we have two magnets what will happen if we put the two north poles together?

e) Which of these things will be magnetic?

a steel nail   a copper pipe   an iron horseshoe
a plastic cup   a nickel coin

Measuring Force

We measure force using a force meter or a newtonmeter.

Newtonmeters are named after the famous scientist Isaac Newton who worked out the scientific laws that explain how objects move. Most newtonmeters and force meters use a spring to measure pull forces.

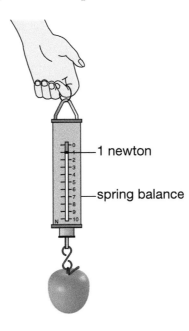

We measure the strength of forces in newtons. We use the symbol N for newtons.

The force of gravity on an object with a mass of one kilogram (1 kg) is almost 10 newtons (10 N). The real value is about 9.8 newtons but we often round this up to 10 when we are working out problems.

So, for example, if a box has a mass of 5 kg we can work out the force of gravity on the box by multiplying the mass by 10.

5 × 10 = 50

The force of gravity on the box is 50 N.

We call the force of gravity on an object the **weight** of the object. So the weight of the box is 50 N. The box has a **mass** of 5 kg and a **weight** of 50 N.

Here is another example. Richard has a mass of 45 kg. The force of gravity is 10 N for each kilogram.

45 × 10 = 450

The force of gravity on Richard is 450 N. So Richard's weight is 450 N.

Exercise 3

a) Which of these units do we use to measure force?

kilometres (km)   grams (g)   tonnes (t)   newtons (N)   metres (m)

b) If the force of gravity on one kilogram is 10 N, what is the force of gravity on a bag that has a mass of 12 kg?

c) What is the weight of the 12 kg bag?

d) Aisha has a mass of 36 kg. What is her weight?

Remember!

◆ Forces are nearly always pushes or pulls.

◆ Forces can make an object do five things: speed up, slow down, change direction, turn or change shape.

◆ If the forces on an object are unbalanced, the object will move in the direction of the strongest force.

◆ Air resistance and friction are two forces that will slow down a moving object.

◆ We can make an object turn by pushing or pulling a lever, such as a door or a spanner, which then turns around a pivot.

◆ Magnetism and gravity are non-contact forces that can move an object without touching it.

◆ We measure forces in newtons (N).

◆ The force of gravity on one kilogram is about 10 N.

Revision Test on Forces

**Now that you have worked your way through the chapter, try this
revision test. The answers are in the answer book.**

1. What units do we use to measure force?
2. Name an instrument that you could use to measure force.
3. Does friction make moving objects slow down or speed up?
4. Why do we need friction if we are cycling or travelling in a bus or a car?
5. Why do racing cars and aeroplanes have a streamlined shape?
6. What is a pivot?
7. Will these two magnets attract each other or repel each other?

8. What force makes parachutists fall towards the ground when they jump
   out of a plane?
9. Why does a parachutist slow down when the parachute opens?
10. A cricket ball has a mass of 0.15 kg. What is the force of gravity on the
    cricket ball? What is the weight of the cricket ball?

6 Energy

What is Energy?

Energy is used every time something does some work. It could be a person running, writing, walking or talking. It could be a machine like a tractor, a car or an aeroplane. It could be an electric motor or an electric light. We need energy to allow work to happen. Without energy, work cannot happen. Energy exists in many forms and can be stored in different ways.

Different Forms of Energy

Here are the most common forms of energy.

Chemical Energy

Chemical energy is stored inside a substance and is released by a chemical reaction. The energy stored in food is chemical energy. When we eat and digest food we break it down into chemicals that provide us with energy.

The energy stored in **fuels** like coal, wood and oil is also chemical energy. The energy in a fuel is changed into heat and light when the fuel is burnt.

chemical energy

The chemical energy stored in a battery can be used to make an electric current, which gives electrical energy.

Electrical Energy

Electrical energy is very useful. Many things in our houses, like lights, fridges and TVs, can do work because of electrical energy. Wherever there is an electric current flowing there is electrical energy. Power stations generate electrical energy or we can get an electric current from the chemical energy stored in a battery.

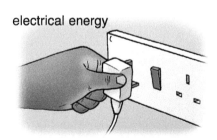

electrical energy

Light Energy

Light is the energy that we detect with our eyes. We get light from the Sun and from electric lights, torches, candles, fires and other hot objects.

light energy

Sound Energy

sound energy

Sound energy is made by vibrating objects. For example, when we speak, the vocal cords in our throat vibrate. When we hit a drum, the drum skin vibrates. When we play a guitar, the strings vibrate.

Kinetic Energy

Kinetic energy is the energy of movement. Any object that is moving has kinetic energy. The faster an object moves, the more kinetic energy it has. This windmill has kinetic energy when it is turning.

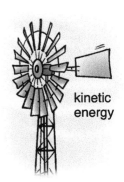

kinetic energy

Gravitational Potential Energy

We learnt in Chapter 5 that gravity pulls an object towards the Earth. If an object is held up above the Earth, for example a book on a shelf or a child at the top of a slide, it has **gravitational potential energy**. This means it has the potential to fall, because of the force of gravity. The energy is released when the object falls. As the object falls, the energy is changed into kinetic energy. In the first drawing below, the child, the book and the plate all have gravitational potential energy. In the second drawing, they are all falling so their gravitational potential energy is being changed into kinetic energy.

Strain Energy

Strain energy is also called **elastic potential energy**. It is the energy stored by stretching or squashing an elastic object. An elastic object is one that will go back to the shape it started with. Stretching a rubber band gives it strain energy. Squashing a spring gives it strain energy.

strain energy

Heat Energy

Heat energy is often called thermal energy. The hotter something is, the more heat energy it has.

heat energy

Exercise 1

a)   What kind of energy does a moving object have?
b)   What kind of energy is stored in a battery?
c)   What kind of energy is stored in stretched rubber band?
d)   What kind of energy does a child at the top of a slide have?
e)   What kind of energy is stored in coal?

Energy Changes

One of the most important things to remember about energy is that the total amount of energy always stays the same. It is never all used up, it just changes from one form of energy to another. So when we use one store of energy, another store of energy increases. In other words, when we do work, energy is changed from one form to another.

For example, when we switch on an electric light, the store of electrical energy is reduced but the amounts of light energy and heat energy are increased. So the total amount of energy stays the same.

Here are some examples of energy changes.

When we switch on an electric light, we are changing electrical energy into light energy and heat energy.

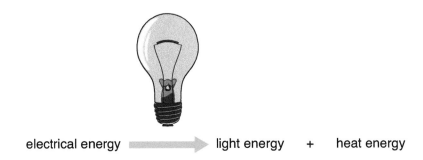

electrical energy ⟹ light energy + heat energy

When we switch on a battery-powered radio, we turn chemical energy in the battery into electrical energy and then into sound energy.

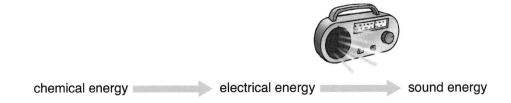

chemical energy ⟹ electrical energy ⟹ sound energy

When the girl dives from the high board into the pool, her gravitational potential energy changes into kinetic energy.

gravitational potential energy ⟹ kinetic energy

If we burn wood in a fire, we are turning chemical energy stored in the wood into heat energy and light energy (and some sound energy too!).

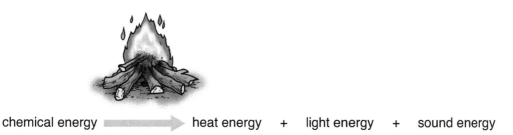

chemical energy ➡ heat energy + light energy + sound energy

Exercise 2

Fill in the gaps in the energy changes below.

a) Brian switches on a battery-powered torch.

_____ → _____ → _____ + _____
   energy   energy   energy   energy

b) Wood is burnt in a fire.

_____ → _____ + _____ + a little
   energy   energy   energy   energy

c) An electric fan is switched on.

_____ → _____
   energy   energy

d) Anne talks to Minesh on the telephone.

_____ → _____ → _____
   energy   energy   energy

e) A child slides down a slide.

_____ → _____
   energy   energy

Energy From the Sun

Most of the energy we use comes from the Sun. The Sun's energy reaches the Earth as heat and light. It is then changed into many other forms of energy.

The diagrams on the next page show some examples.

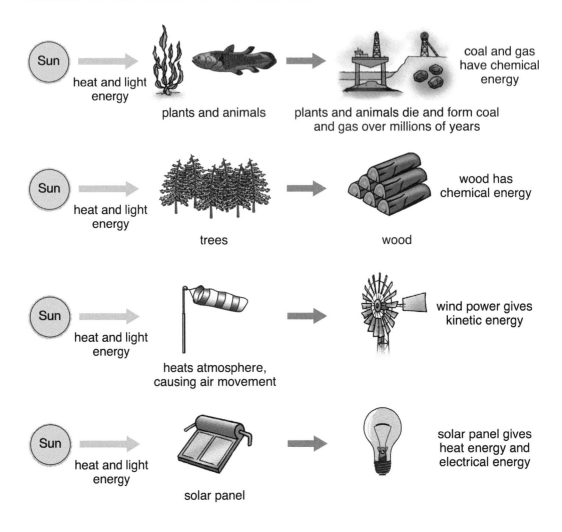

Renewable and Non-renewable Sources of Energy

We can split our sources of energy into two groups:

◆ sources of energy that will run out if we keep using them
◆ sources of energy that won't run out.

We call the sources of energy that will run out **non-renewable** sources. For example, we burn coal, gas and oil to change their chemical energy into heat and light, but all of them are non-renewable because they aren't automatically replaced when we use them up. Coal, gas and oil are all made from dead plants and animals. It takes millions of years for the dead plants and animals to be changed into coal, gas and oil, so we can't replace them if they run out. At the moment people around the world burn a lot of coal, gas and oil to make electricity in power stations and to give heat for cooking and warmth.

We call sources of energy that won't run out **renewable** sources. The Sun is the source of wind, wave and solar power. When we use these sources of energy they are automatically replaced when the Sun shines. The Sun also causes trees to grow, which give us wood to burn. When we use wood we can replace it by planting more trees.

A Closer Look at Heat Energy

Heat is one of the most useful forms of energy. There is a difference between temperature and heat. Temperature is a measure of how hot something is. It is usually measured in degrees Celsius (°C).

Water freezes at 0°C. This is the **freezing point** of water.

Water boils at 100°C. This is the **boiling point** of water.

Heat is a form of energy and shows how much heat something has altogether. For example, a small cup of water at 60°C has less heat energy altogether than a large bath at 50°C because the bath has so much more hot water.

Solids, Liquids and Gases

Heat energy helps us to explain why **solids**, **liquids** and **gases** are different. If a solid gains enough heat energy its temperature will rise and it will melt and become a liquid. If a liquid gains enough heat energy it will become a gas.

We can think of things as being made of tiny particles.

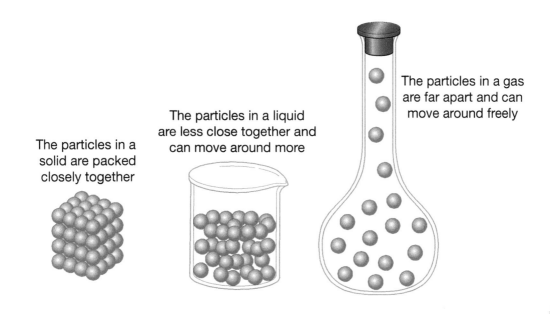

The particles in a gas are far apart and can move around freely

The particles in a liquid are less close together and can move around more

The particles in a solid are packed closely together

If we heat a solid, the heat energy gives the particles more and more kinetic energy until the solid melts and becomes a liquid. The particles now move around faster and more freely. If we keep heating the liquid, the heat energy keeps giving the particles more and more kinetic energy until the liquid **evaporates** and becomes a gas. The particles are now far apart and are moving even faster.

Heat Transfer

Heat energy flows between things that have different temperatures. Heat energy always flows from a hotter thing to a colder thing.

heat flows from hot to cold

For example, a hot cup of tea gets colder because the heat energy flows from the hot cup to the cooler air around it.
If we wait long enough the tea will cool down to the same temperature as the air around it.

A cold can of drink taken out of a fridge will get warmer because heat energy flows from the warmer air to the colder can. If we wait long enough the can will warm up to the same temperature as the air around it.

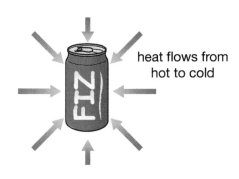

heat flows from hot to cold

When heat flows from a hot thing to a cold thing we call it **heat transfer**.

Heat can be transferred in three ways – by conduction, convection and radiation.

Conduction

Conduction is when heat energy spreads through an object from the hotter part to the cooler part. For example, if we put a metal poker in a fire, the heat energy will be transferred by conduction from the hot end to the cooler end of the poker. The heat energy is passed from particle to particle as it spreads through the solid.

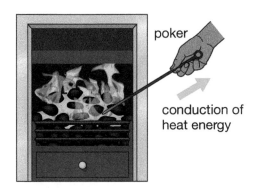

poker

conduction of heat energy

A solid object that allows heat energy to flow through it easily is a good conductor. Most metals are good conductors. Materials like wood and plastic are bad conductors. We call things that conduct heat badly **insulators**. We use insulators to stop heat being transferred from one object to another. For example, metal cooking pots and pans often have wooden or plastic handles. The cooking pot is made of metal because it is a good **conductor** of heat. The handle is made of wood or plastic because it is a good insulator and stops the heat from being transferred from the cooking pot and burning our hands.

Convection

Convection is when heat energy spreads through a liquid or gas. In convection the particles in the gas or liquid move from a hot place to a cooler place in the gas or liquid. As they move, cooler particles take their place. These particles are then heated by the source of the heat energy. Liquids and gases can transfer heat by convection because the particles in liquids and gases can move around, unlike the particles in a solid.

Here is an example of a classroom being heated by convection. The hot air particles rise above the heater and are replaced by cooler air particles. In this way heat is spread through the room.

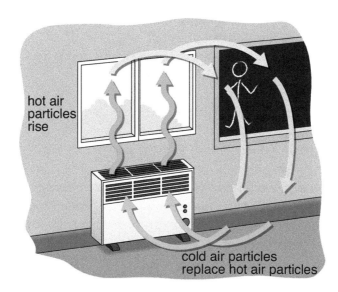

hot air particles rise

cold air particles replace hot air particles

Radiation

Radiation is when heat energy is spread without using particles. For example, heat travels from the Sun to the Earth. There are no particles in space, so the heat travels by radiation. We call an empty space with no particles in it a **vacuum**.

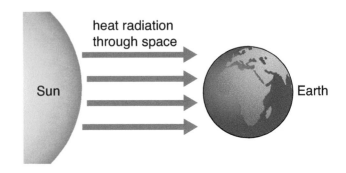

Exercise 3

a) Give two examples of non-renewable sources of energy.
b) Give two examples of renewable sources of energy.
c) Fill in the missing word:
   Heat energy can flow between two things if there is a difference in

   _____.

d) Give an example of a material that is a good conductor.
e) Give an example of a material that is a good insulator.
f) What is radiation?

Remember!

◆ Energy is used every time something does some work.
◆ There are many different forms of energy.
◆ The total amount of energy always stays the same, but it gets changed from one form into another when work is done.
◆ Most of the energy we use started in the Sun.
◆ Non-renewable sources of energy, such as coal, gas and oil, will run out if we keep using them.
◆ Renewable sources of energy, such as solar power and wind power, are replaced when the Sun shines.
◆ Heat energy flows between things that have different temperatures.
◆ Heat energy always flows from hot to cold. This is called heat transfer.
◆ There are three ways that heat energy can be transferred – conduction, convection and radiation.

Revision Test on Energy

Now that you have worked your way through the chapter, try this revision test. The answers are in the answer book.

1. Name five forms of energy and give an example of each one.
2. Which energy change happens when we switch on a battery-powered radio?
   - ◆ electrical energy → heat energy → light energy
   - ◆ chemical energy → gravitational potential energy
   - ◆ chemical energy → electrical energy → sound energy
   - ◆ kinetic energy → sound energy
3. Which energy change happens when we switch on an electric light?
   - ◆ electrical energy → sound energy
   - ◆ electrical energy → kinetic energy + heat energy
   - ◆ gravitational potential energy → kinetic energy + light energy
   - ◆ electrical energy → heat energy + light energy
4. Why is coal a non-renewable energy source?
5. Why is solar power a renewable energy source?
6. Here is a drawing of the particles in three substances.
   Which substance is a solid, which is a liquid and which is a gas?

   A   B   C

7. When heat spreads through a solid object from hot to cold it is called:

   radiation conduction convection

8. When heat spreads through a liquid or gas from hot to cold it is called:

   radiation conduction convection

9. When a saucepan of water is heated on a stove, the hot water particles rise from the bottom of the pan and are replaced by cooler water particles. In this way heat is spread through the water. This is an example of:

   radiation conduction convection

10. Give an example of a material that is a good conductor of heat.
11. Give an example of a material that is a good heat insulator.
12. Should the handle on a kettle be made from a good insulator or a good conductor?

7 Light and Sound

What is Light?

We learnt in the last chapter that we detect light with our eyes. We get light from the Sun and from electric lights, torches, candles, fires and other hot objects. Objects that give out light are called **luminous** objects.

Objects that don't give out light are called **non-luminous** objects. Most objects are non-luminous. Look around you. How many objects can you see that are luminous and how many that are non-luminous?

The Speed of Light

Light travels in a straight line at a very high speed. Light travels at a speed of about 300 000 000 metres every second. That's just over 1 000 000 000 kilometres per hour!

How Do We See Things?

We can see luminous objects because the light rays they give out travel to our eyes.

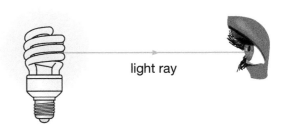

light ray

We can see non-luminous objects only if there is a luminous object such as the Sun or an electric light to **illuminate** them (shine light on them). The light from the luminous object bounces off the non-luminous object towards our eyes. When light bounces off an object we call it **reflection**. When this happens we can see the non-luminous object. For example, imagine a dark room with a table in it. You can't see the table until the electric light is switched on. You can then see the table because the light from the electric light bounces off the table to your eyes.

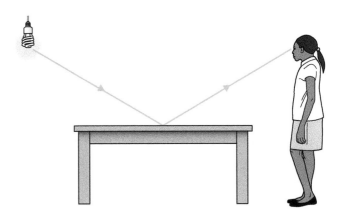

Transparent, Translucent and Opaque

Light doesn't always bounce off things. Light can also travel through materials such as glass, water and clear plastics. Because light travels through these materials we can see through them. Materials we can see through are called **transparent** materials. When light travels through them we say the light has been **transmitted**. Clear glass is an example of a transparent material.

Some materials let some light pass through them but we can't see clearly through them. We call these materials **translucent**. Frosted glass is an example of a translucent material.

Materials we can't see through at all are called **opaque** materials. A brick wall is an example of an opaque material.

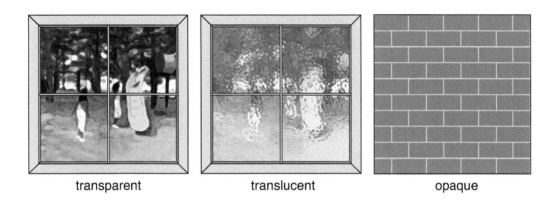

transparent   translucent   opaque

Because light rays can't travel through an opaque material, when the Sun shines on it there will be a **shadow** where the light rays cannot reach. Notice that when the Sun is low the shadow is longer than when the Sun is high. This is because the light rays travel in straight lines.

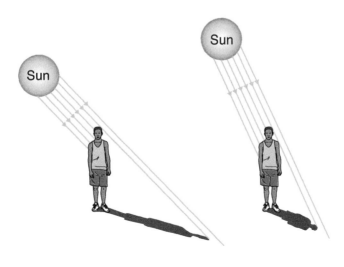

Exercise 1

a) What is a luminous object? Give an example.
b) What is a non-luminous object? Give an example.
c) How can we see a non-luminous object?
d) What is a transparent material? Give an example.
e) What is an opaque material? Give an example.

A Closer Look at Reflection

We have just learnt that when light bounces off something, it is called **reflection**. Light bounces well off mirrors and shiny objects. We say that mirrors and shiny objects **reflect** light well. Dull, dark and black objects don't reflect light well. They **absorb** a lot of the light.

The angle at which a light ray is reflected from a mirror is always the same as the angle at which the ray arrived at the mirror.

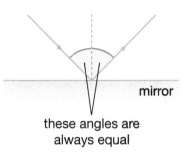

By changing the angle between the mirror and the light rays we can reflect light in any direction.

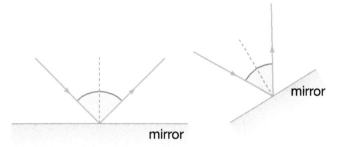

When we look at something in a mirror, the **image** we see is always the opposite way round to the real object. This is because of the way the light rays from the object are reflected. In this drawing we can see that the ray from the right side of the object becomes the ray on the left side when it is reflected.

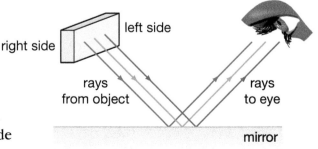

We can see an image in a mirror because it is smooth and reflects light rays very well. If we look at a rough surface like a wall we can't see a reflected image. This is because the rough surface means the reflected light rays are scattered in different directions so we don't see an image.

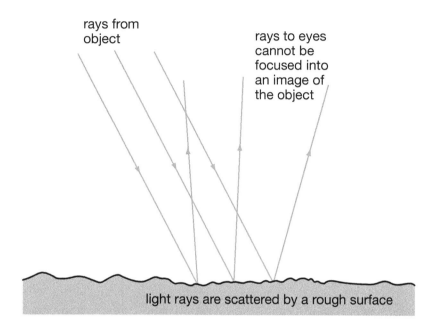

rays from object

rays to eyes cannot be focused into an image of the object

light rays are scattered by a rough surface

What is Refraction?

When light passes through transparent materials like water, glass or clear plastic, it is not reflected. The light rays bend a little as they move from the air into the different material, and this makes them change direction slightly.

For example, if we put a straw in a glass of water it looks bent.

The straw isn't really bent, it's just that the light rays have changed direction slightly when they pass from the air to the water and back to the air again.

When lights rays change direction like this it is called **refraction**.

We use refraction in specially shaped pieces of glass or transparent plastic called **lenses**. Lenses are shaped so that they refract light in different ways. The most common types of lenses are called **convex** and **concave**. Convex lenses refract light rays towards a central point called the **focal point**. Concave lenses refract light rays so that the light rays spread out.

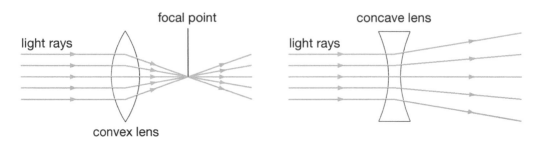

Light and Colour

The light we see from the Sun is called **white light**. White light is made up of many different colours. More than 300 years ago, Isaac Newton was the first scientist to discover that white light is made up of different colours, when he passed a beam of sunlight through specially shaped pieces of glass called **prisms**. The white light was split into seven colours: red, orange, yellow, green, blue, indigo and violet light.

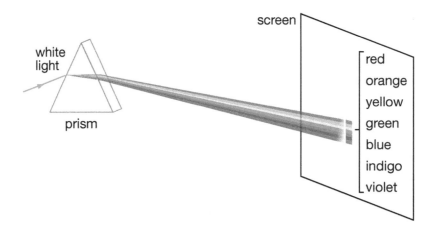

When the white light passes through the prism, the light of each colour is refracted by a slightly different amount. Red light is refracted the smallest amount and violet light is refracted the most. We call the range of colours the **spectrum** of light.

How Are Rainbows Made?

If you stand with your back to the Sun when it is raining you might see a rainbow. Rainbows are made when light from the Sun is refracted and then reflected by raindrops. First the light is refracted by the raindrops and then it is reflected to your eye. The drawing on the next page shows this.

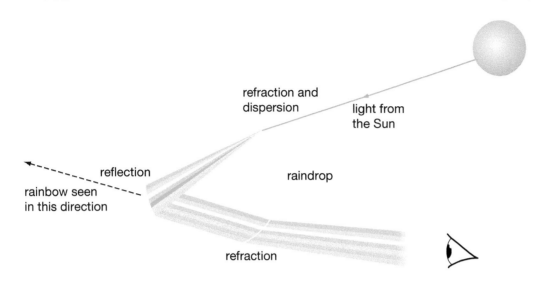

Why Do Different Things Have Different Colours?

We learnt at the start of this chapter that we see things because they reflect light to our eyes. The colour of an object is caused by the light it reflects. For example, a red object absorbs all the colours in white light apart from red. The red light is reflected to your eye so the object looks red. A green object absorbs all the colours in white light apart from green. The green light is reflected so the object looks green.

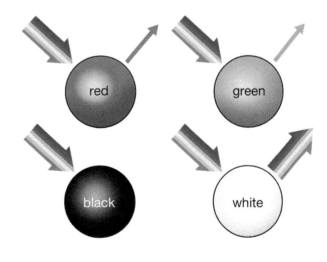

A white object reflects all the colours and so the object looks white. A black object absorbs all of the colours and reflects none, so the object looks black.

Exercise 2

a) Fill in the gaps in these sentences.

When light bounces off an object it is called _____. When light moves from one transparent substance to another it changes direction slightly. This is called _____.

Exercise 2 *(continued)*

b) Which of these objects are good reflectors of light?

a piece of coal   a mirror   a shiny plate   a dark blue shirt

c) On this picture draw in a light ray to show how the light is reflected by the mirror.

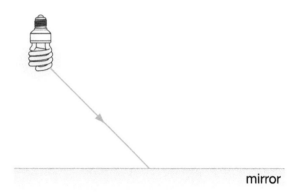

mirror

d) What are the colours in the spectrum of white light?
e) Why does a blue object look blue?

Our Eyes

Here is a drawing of a human eye, seen from the front, and another drawing showing the structure of the eye.

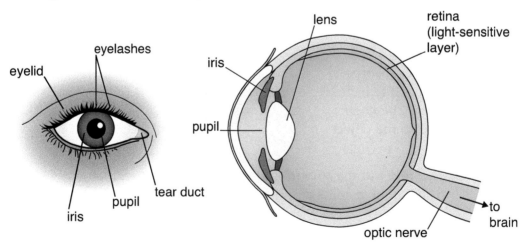

Light rays pass through the pupil, which is the gap in the middle of the iris. The light rays are then refracted by the lens before they go to the retina at the back of the eye. An image of what the eye sees is made on the retina. The retina is made up of special light-sensitive cells and so information about this image is then sent along the optic nerve to the brain.

Some people have problems seeing objects clearly because the light rays don't make an image properly on the retina.

A **short-sighted** person can see near objects clearly, but objects far away cannot be seen clearly. This is because the lens in their eye refracts the light too much and so the image of a far-away object is made before it reaches the retina.

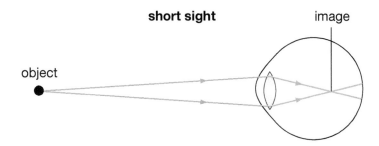

Short-sighted people can see far away objects clearly by wearing spectacles (glasses) with concave lenses. The concave lens in the spectacles spreads the light rays out before they enter the eye. The lens in the eye then refracts the light rays onto the retina.

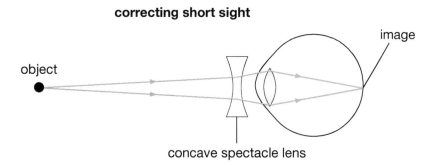

A **long-sighted** person can see distant objects clearly, but they cannot see near objects clearly. Their eyes do not refract light enough for the image of the object to be made on the retina.

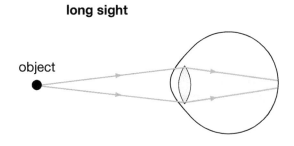

Long-sighted people can see near objects by wearing spectacles with convex lenses. The convex lens in the spectacles refracts the light rays inwards before they enter the eye. The lens in the eye then refracts the lights rays onto the retina.

correcting long sight

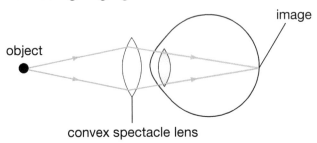

object

image

convex spectacle lens

Exercise 3

a) What name do we give to the gap in the iris which lets light into the eye?
b) An image of what the eye sees is made on the back of the eye. What is this part of the eye called?
c) What does the optic nerve do?
d) What type of lens should a short-sighted person have in their spectacles?
e) What type of lens should a long-sighted person have in their spectacles?

What is Sound?

We learnt in the last chapter that sounds are made by vibrating objects. For example, when we speak, the vocal cords in our throat vibrate. When we hit a drum, the drum skin vibrates. When we ring a bell, the bell vibrates.

When an object vibrates, the air around it also vibrates. This vibration then passes through the air in a **wave**. This is how sound travels.

Sound can also travel through solid materials such as metal, wood and stone. In each case, the vibrations from the object that makes the sound cause the particles in the solid to vibrate too.

Sound can also pass through water and other liquids, but it cannot pass through a vacuum. In the last chapter we learnt that a vacuum is a completely empty space with no particles in it at all, not even air particles. Sound cannot travel through a vacuum because there are no particles to vibrate in a vacuum.

Sound travels through air at a speed of about 340 metres per second. This means that sound is *much* slower than light. In a thunder storm the lightning and the thunder happen at the same time, but because light travels much faster than sound we see the lightning before we hear the thunder.

Loudness and Pitch

The more energy there is in a vibration, the louder the sound will be. This means that the harder we hit a drum or pluck a guitar string the louder the sound will be, because we have given the vibration more energy. Loudness is measured on a scale called the **decibel** scale. Decibels have the symbol dB. This table gives the loudness of some ordinary sounds in decibels.

Source of sound	Loudness in decibels (dB)
quietest sound that humans can hear	0
whisper	20
normal talking	60
heavy traffic	90
jet aeroplane taking off	140

The **pitch** of a sound is how high or low the sound is. Short vibrating objects give higher sounds than long vibrating objects. For example, a short guitar string will give a higher note than a long guitar string, and a small drum will give a higher note than a big drum. Some animals can hear much higher notes than humans can. For example, dogs, dolphins and bats can all hear much higher notes than humans.

The large bass drum makes a much lower sound than the small snare drum.

Echoes

When sound travels through something like air or water we say the sound is transmitted. Sound can also be reflected and absorbed. For example, if you stand in a big empty room and speak or clap your hands you will hear an **echo**. This is sound being reflected from the hard surfaces around the room.

If you speak or clap in a room with lots of soft materials such as curtains or carpets you won't hear such a loud echo. This is because the sound is absorbed by the soft materials.

Our Ears

Here is a drawing of a human ear. The ear is made up of three main parts, called the outer ear, the middle ear and the inner ear.

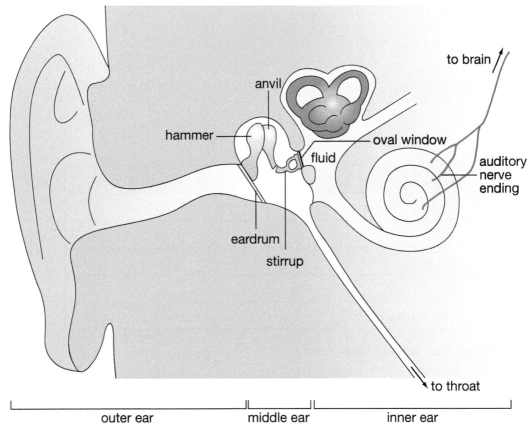

outer ear   middle ear   inner ear

We hear sounds when the vibrating air, or sound wave, reaches our ear and passes through the outer ear to the eardrum. The vibrations make the ear drum vibrate, which makes the three bones in the middle ear vibrate. These three bones are called the hammer, the anvil and the stirrup, and their job is to make the vibration stronger. The vibration is then passed to the inner ear. The auditory nerve in the inner ear then sends a message to our brain. The brain understands the message from the auditory nerve and works out what sounds we are hearing.

Very loud sounds can damage our ears. If we listen to sounds of more than 90 dB for a long time our ears will be damaged. Very loud sounds like a jet aeroplane taking off can damage our hearing very quickly if we don't wear ear protection.

Exercise 4

a)   What causes sound?
b)   Which travels faster – light or sound?
c)   What is the scale we use to measure loudness?
d)   Which gives a lower note – a long guitar string or a short guitar string?
e)   How does information about sounds get from our ear to our brain?

Remember!

- Objects that give out light are called luminous objects. Objects that don't give out light are called non-luminous objects.
- Light travels in straight lines at very high speed.
- When light reaches an object it is transmitted (which means the light passes through the object), reflected or absorbed.
- Transparent materials are materials we can see through. Translucent materials let some light through but not enough to see through them clearly. Opaque materials don't let any light through so we can't see through them.
- Mirrors are good reflectors of light. Dull, dark and black objects are poor reflectors of light.
- The angle at which a light ray is reflected from a mirror is always the same as the angle at which the ray arrived at the mirror.
- Light rays change direction slightly when they pass from one transparent material to another. This is called refraction.
- White light is made up of seven different colours: red, orange, yellow, green, blue, indigo and violet. These colours are called the spectrum of light.
- Objects have different colours because they absorb some colours of light and reflect others.
- Light rays enter the human eye through the pupil. They are then refracted by the lens to form an image on the retina. Information about the image is then sent along the optic nerve to the brain.
- Sounds are made by vibrating objects.
- The more energy there is in a vibration, the louder the sound. We measure loudness in decibels.
- The pitch of a sound is how high or low it is.
- Short vibrating objects give higher sounds than long vibrating objects.
- When sound is reflected we hear an echo.
- We hear sounds when the vibrating air reaches our ear and passes through the outer ear to the eardrum. The vibration is then passed on through the middle ear to the inner ear and then a message is sent along the optic nerve to the brain.

Revision Test — on Light and Sound

Now that you have worked through the chapter, try this revision test. The answers are in the answer book.

1. Daniel is reading a book. Draw a light ray on the picture to show how Daniel can see the book.

2. What does 'opaque' mean?
3. What does 'transparent' mean?
4. Which two of these four objects are the best reflectors of light?

   a black wall   a white wall   a brown coat   a yellow shirt

5. What are the different colours in the spectrum of white light?
6. Light of only one colour is reflected by a red object. Which colour is it? What happens to the other colours?
7. When light passes from air to water it changes direction slightly. What is this change of direction called?
8. Here is a drawing of an eye. Fill in the missing labels

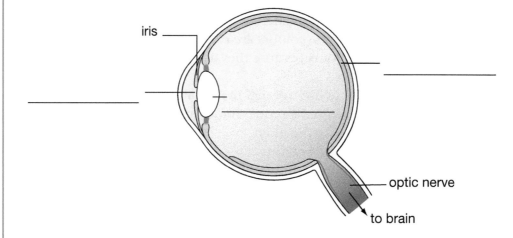

iris

optic nerve

to brain

9. In a thunder storm, lightning and thunder happen at the same time. Why do we see the lightning before we hear the thunder?

Revision Test | on Light and Sound *(continued)*

10. The drawing shows a small string instrument called a violin and a larger string instrument called a cello. Which instrument will give notes of a higher pitch?

violin

cello

11. Which of these four things explains what an echo is?

refraction of sound   absorption of sound
transmission of sound   reflection of sound

12. Here is a drawing of an ear. Fill in the missing labels.

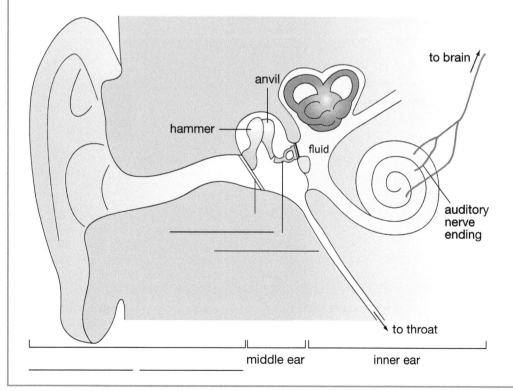

to brain

anvil

hammer

fluid

auditory
nerve
ending

to throat

middle ear   inner ear

8 Electricity

In Chapter 6 we learnt that electricity is a very useful form of energy. Many things in our houses, such as lights, refrigerators and TVs, can do work because of electrical energy. Wherever there is an electric current flowing there is electrical energy. Power stations generate electrical energy, or we can get an electric current from the chemical energy stored in a battery. The electricity from power stations that we use in our buildings and houses is called **mains** electricity.

Electric Current

When you switch on an electrical device such as a light, a TV or a computer, electricity flows from the power source along the wires to the device and then back again to the power source. For small devices we often use batteries as the power source. The electricity that flows along the wires is called an **electric current**.

In this drawing, the electric current flows from the battery to the bulb and then back to the battery.

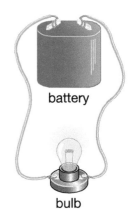

battery

bulb

Electric Circuits

The wires and other **components** (like bulbs) that the current flows through together make an **electric circuit**. A circuit is something that ends up back where it started. For an electric current to flow, the circuit has to be joined up all the way back to the power source. If there is a gap in the circuit then the electric current will not flow around the circuit.

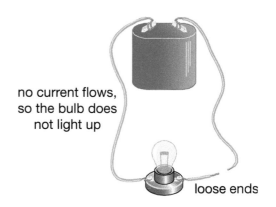

no current flows, so the bulb does not light up

loose ends

We can make simple electric circuits using a battery and components. The battery is the power source and the components are things that go in the circuit, such as a light bulb, an electric bell or an electric motor. The battery and the components must be connected with wires so that there is no gap in the circuit.

Circuit Diagrams

When we draw a picture of a circuit, we call it a **circuit diagram**. Each of the components in a circuit diagram has a symbol. Here are the most common symbols.

Component	Picture	Symbol
battery		
two batteries		
bulb		
buzzer		
motor		
switch – off		
switch – on		

Here is a circuit with a switch and a bulb, and the circuit diagram that goes with it.

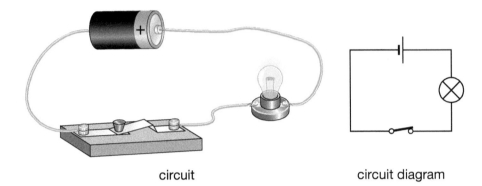

circuit   circuit diagram

Switches are very useful components because they let us stop the flow of electricity around the circuit. Switches save money and power from the power source because they let us switch things off.

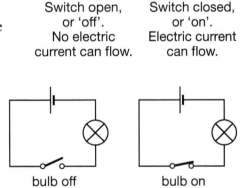

Switch open, or 'off'. No electric current can flow.

Switch closed, or 'on'. Electric current can flow.

In this simple circuit, the bulb will light when the switch is closed.

bulb off   bulb on

If we add more batteries in a line to the circuit, the bulb gets brighter. The more batteries we add, the brighter the bulb will be.

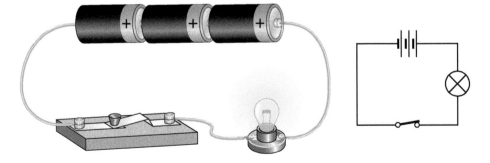

Instead of adding more batteries, we could add more bulbs to the circuit. If we do this, the bulbs will be dimmer than a single bulb. The more bulbs we add, the dimmer the light will be. The drawings on the next page show what happens in a circuit with one battery and three bulbs.

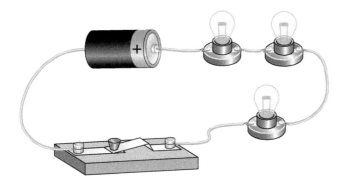

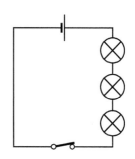

If we open the switch in any of the circuits, the bulbs will go out because no electric current will flow.

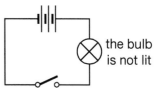

the bulb is not lit

Exercise 1

a) Give the names of the components labelled A to D in this circuit diagram.

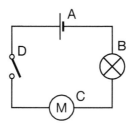

b) Look at these circuit diagrams. Which of the bulbs will light?

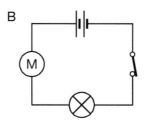

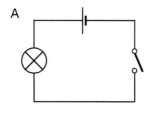

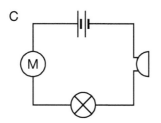

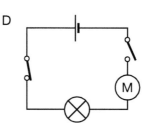

Conductors and Insulators

Materials that allow electric current to flow through them are called **conductors**. Metals are the best conductors as they allow electric current to flow through them very easily. The wires in an electric circuit are made of metal. Copper is the metal most often used for wires in a circuit.

Materials that don't allow electric current to flow through them are called **insulators**. Materials such as plastic, rubber, glass and wood are all good insulators. The metal wires in an electric circuit are usually wrapped in plastic. Because plastic is an insulator, we can pick up the wire without getting an electric shock from the electric current flowing through the wire.

Safety

Electricity is very useful but it can also be dangerous. The electricity that flows through the mains in a building is a lot more powerful than electricity from a battery. An electric shock from the mains could kill you, so:

◆ Never stick anything other than a proper electrical plug into a mains socket.
◆ Never hold the metal part of an electrical plug when you are plugging it in.
◆ Never touch switches with wet hands.
◆ Never use electrical machines or devices close to water.

Exercise 2

Here are the names of five materials

aluminium   wood   glass   rubber   iron

a) Which of these materials are good conductors?
b) Which of these materials are good insulators?
c) Which allow electric current to flow through them – the conductors or the insulators?

Remember!

◆ Electric current flows around an electric circuit.
◆ An electric circuit must be joined up all the way round, starting at the power source and ending back at the power source.
◆ If there is a gap in an electric circuit then electric current won't flow.
◆ When we draw a picture of a circuit, we call it a circuit diagram.
◆ Each of the components in a circuit diagram has a symbol.
◆ Materials that allow electric current to flow through them are called conductors.

Remember! *(continued)*

◆ Materials that don't allow electric current to flow through them are called insulators.
◆ Metals such as copper are the best conductors.
◆ Plastic, rubber, glass and wood are all examples of good insulators.
◆ Electricity is very useful but it can be dangerous. We must be very careful when using electricity from the mains as it is much more powerful than electricity from batteries.

Revision Test | on Electricity

Now that you have worked through the chapter, try this revision test. The answers are in the answer book.

1. Name five things that need electricity to work.
2. Why are the wires in an electric circuit made of metal?
3. Why is there usually a plastic coating around the metal wires in an electric circuit?
4. Draw the circuit symbols for each of these components:

   bulb   battery   open switch   closed switch   electric motor

5. Draw a circuit diagram showing a circuit that has a battery, a bulb and an open switch.
6. If you add an extra battery to the circuit you drew for question 5, what will happen to the brightness of the bulb?
7. If you add an extra bulb to the circuit you drew for question 5, what will happen to the brightness of the bulbs?
8. Which of these bulbs will light?

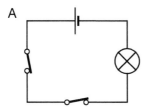
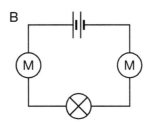
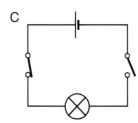

9. Why is it useful to have switches in electric circuits?
10. Which of these materials are good insulators?

   plastic   copper   steel   aluminium   glass

11. Explain why electricians wear thick rubber gloves when they are working on electric wires.

9 Rocks, Minerals and Soil

Rocks, **minerals** and **soil** are all around us. The surface of the Earth is made from rocks. People have used rocks for thousands of years to build homes and to make tools. Most rocks are made from a mixture of minerals. Minerals are natural substances found in the ground that join together to make rocks. Soil is made from tiny pieces of rock that have broken off from bigger rocks.

Rocks

Not all rocks are the same. Some are harder than others. Some are very soft and can be scratched or easily crumbled. Some rocks are shiny and some are dull. Soft rocks often let water soak through them but hard rocks don't usually let water soak through.

Some rocks are a mixture of other smaller rocks, and some rocks have the shapes of long-dead plants and animals in them. These shapes are called **fossils**. Rocks that are a mixture of other rocks, and rocks that have fossils in them, usually feel rough.

Rocks are different to each other because they are made, or formed, in different ways. There are three main kinds of rock, which are called **igneous** rocks, **sedimentary** rocks and **metamorphic** rocks. Each is formed in a different way.

Igneous Rocks

Inside the Earth it is very hot. A very hot runny material called **molten magma** forms inside the Earth. This material rises up to the Earth's surface where it cools down. As it cools down it turns to solid rock. Rocks made like this are called igneous rocks.

Sometimes molten magma is thrown out by an erupting volcano. The magma forms lava as it runs down the side of the volcano. The lava cools down to form igneous rocks.

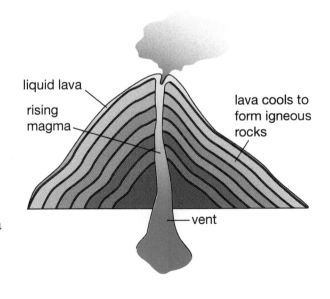

liquid lava

rising magma

lava cools to form igneous rocks

vent

Crystals are formed in the rock as it cools. If the magma cools slowly the crystals are large and easy to see. If the magma cools quickly the crystals are small and difficult to see.

Granite is a common igneous rock. Granite is formed when magma cools down slowly, so it has large crystals in it.

Other common igneous rocks are **basalt**, **obsidian** and **pumice**. These rocks are all made when magma or lava cools down quickly, so they have very small crystals or no crystals at all.

granite, with crystals

Sedimentary Rocks

Broken bits of rock and pieces of dead plants and animals are washed into the sea and into lakes by rivers and by rain. These bits settle at the bottom of the sea and lakes and form layers. These layers are called **sediments**. The layers of sediments become squashed over millions of years by other layers that form on top of them. As the layers are squashed they harden into rock. Rocks formed in this way are called sedimentary rocks.

Sedimentary rocks are often a mixture of other rocks and some have fossils in them. Sedimentary rocks are often soft, such as chalk. Other types of sedimentary rock are sandstone, limestone, conglomerate (which is a mixture of bits of other rocks) and shale.

conglomerate, containing bits of rock

Sedimentary rocks can also be formed when a sea dries up and leaves crystals behind. Rock salt is a sedimentary rock formed in this way.

Metamorphic Rocks

Metamorphic rocks are made from igneous or sedimentary rocks that have been squashed or heated inside the Earth. The squashing and the heating make the rocks change their form into a new rock. The word 'metamorphosis' means 'change'. Shale is changed into slate. Limestone is changed into marble, and granite is changed into gneiss (which we say as 'nice').

gneiss, with striped bands

What Do We Use Rocks For?

Uses of Igneous Rocks

Igneous rocks such as granite and basalt are very hard and so are used for building and for paving roads. Obsidian is an igneous rock that can be broken to give a very sharp edge. In the Stone Age, people used obsidian for knives and arrow heads.

Uses of Sedimentary Rocks

Sedimentary rocks such as sandstone and limestone are more easily dug out of quarries than some harder rocks, and they are easily broken into smaller rocks that can be used for building. Limestone and rock salt are also used a lot in the chemical industry.

Uses of Metamorphic Rocks

Metamorphic rocks such as marble have an attractive appearance, with colourful streaks of minerals. Because it looks good, marble is often used as a building material and also for monuments and statues. Slate is a metamorphic rock that can be split into lightweight sheets. It is also very good at keeping out water so it is a very useful material to use for roof tiles.

Exercise 1

a) Match each rock type with the way it is formed:

igneous rock — formed from layers of bits of rock and pieces of dead plants and animals

sedimentary rock — formed when other rocks are squashed or heated in the ground

metamorphic rock — formed when molten magma from inside the Earth cools down

b) Use these words to fill in the spaces in the sentences below:

sedimentary   igneous   layers   hard   soft   metamorphic   granite

Basalt and _____ are examples of _____ rocks. They are very _____ rocks. Sedimentary rocks form in _____ and are often very _____. Chalk is an example of a soft _____ rock. Marble is a _____ rock that is often used for monuments and buildings.

c) Give one common use of igneous rocks and one common use of sedimentary rocks.

Minerals

Most rocks are made from a mixture of minerals. Minerals are natural substances that are found in the ground that join together to make rocks. Most minerals are made up from two or more chemicals that join together to form a crystal, but a few minerals are made up of just one chemical element. Gold, silver and copper are examples of chemical elements that are found on their own.

Some minerals are shiny and some are dull. Some are soft and some are hard. Diamonds are the hardest mineral and so they are often used for the cutting edges of drills, as well as in jewellery.

Soil

We have learnt that the surface of the Earth is mostly made of rocks, but when we stand on the Earth's surface we are usually standing on soil. Soil is needed for most plants to grow, but where does it come from? Soil is mostly made up of tiny pieces of rock that have broken off from bigger rocks. When a rock breaks up to form smaller pieces it is known as **weathering**. There are two kinds of weathering: **physical weathering** and **chemical weathering**.

Physical Weathering

Physical weathering is a very slow process that slowly breaks down rocks over a long time. There are four main causes of physical weathering:

◆ **Changes in temperature**. Hot weather causes rocks to expand slightly, which means they get a little bit bigger. When the rock cools down at night it shrinks again, which can make the rock crack. This makes small pieces fall off.

Day time

expansion

Heat causes solid rock to expand. The outside layers expand most.

Night time

Cooling causes the rock to contract.

Small pieces break off.

◆ **Freezing and thawing of ice**. Rain can get into small cracks in rocks and in cold countries the rain water can freeze to form ice. Ice takes up more room than rain water so it expands, which makes the rock crack. When the ice melts it forms water again, which can then get into more cracks.

Rain   Freezing conditions

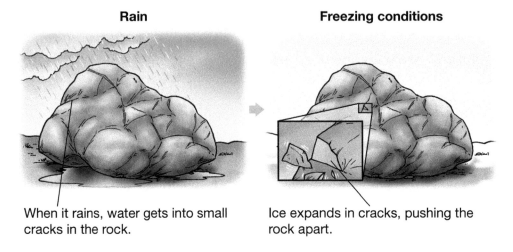

When it rains, water gets into small cracks in the rock.

Ice expands in cracks, pushing the rock apart.

◆ **Rocks rubbing together**. When rocks rub together small pieces break off. For example, pebbles on a river bed rub against each other. Sand grains are tiny fragments of rock and when they blow in the wind they rub on rock surfaces and wear them down.

◆ **Plant roots**. Plant roots can grow in the cracks in rocks. As the plant roots become bigger and stronger they push on the sides of the crack and bits of rock break off.

Chemical Weathering

Chemical weathering is also a slow process. There are two main causes of chemical weathering:

◆ **Rain water.** Carbon dioxide in the air dissolves in rain water. This makes the rain water slightly acidic. The acid in the rain water falling on rocks can react with the rocks and break them up into smaller pieces.

◆ **Hot and wet weather**. The oxygen in the air and dissolved in rain water can also react with rocks to break them up. The reaction of oxygen with rocks works faster in hot and wet weather. Because of this, rocks break down more quickly in hot and wet places such as rainforests than in cold and dry places like the Antarctic.

The tiny pieces of rock that are broken off by weathering are carried away by water or by the wind. When the water or wind slows down, the tiny pieces settle on the ground and can become soil.

Types of Soil

Soil is mostly made of small pieces of rock, but it also has air, water and material from dead plants and animals in it. So soil is made of four things altogether:

◆ small pieces of rock
◆ **humus**, which is made from dead and rotting plants and animals
◆ air
◆ water.

There are different kinds of soil because there are different kinds of rock. The different kinds of soil have different properties. Three of the most common kinds of soil are sandy soil, clay soil and loam.

Sandy Soil

Sandy soil is light and easy to dig. It has lots of air gaps in it so water drains through quickly. There are not many minerals in sandy soil that are useful for plants, and not much water, so plants don't grow well in sandy soil.

Clay Soil

Clay soil is heavy and doesn't allow water to drain through it easily. It is very sticky when it is wet. It is better than sandy soil for growing plants because it has more water and more useful minerals.

Loam

Loam is a very good soil for growing plants. It is often called garden soil. It is made from a mixture of clay and sand and also has a lot of humus in it. Loam holds water and there are a lot of useful minerals in it for plants.

Soil Erosion

Good soil can be blown away by the wind or washed away by water. When this happens we call it **soil erosion**. People can cause soil erosion by not looking after the land. Three common ways that people cause soil erosion are:

◆ **Cutting down trees**. Tree roots help to hold the soil together. This is very important on slopes and hillsides where water can easily wash away soil. Also, the leaves from the trees drop to the ground and form humus, which is needed for good soil.
◆ **Keeping too many animals in a field**. If there are too many animals on a patch of land they will eat all the grass and kill other plants. This is called **over-grazing**. The land becomes bare ground, which is easily eroded by wind and rain.
◆ **Planting the wrong crops**. Different plants grow better in different types of soil. Planting the wrong plants can damage the soil.

Soil Conservation

It is very important to protect soil so that we can grow plants in it. When we protect soil we call it **soil conservation**.

Here are five ways to help soil conservation:

◆ **Planting trees on slopes and hillsides**.
◆ **Not having too many animals on the land**.
◆ **Planting the right crops for the soil**.
◆ **Planting different crops each year**. Different plants take up different minerals from the soil. By planting different crops each year we can slow the loss of some minerals. This is called **crop rotation**.
◆ **Terracing the land on steep hillsides into flat steps**. The flat steps stop rain water from washing soil down the hill.

Exercise 2

a) Soil is made from four things. What are they?
b) Give one example of physical weathering of rocks.
c) Give one example of chemical weathering of rocks.
d) Name three kinds of soil.
e) Give two ways that people can cause soil erosion.

Remember!

◆ The three main types of rock are **igneous** rocks, **sedimentary** rocks and **metamorphic** rocks.
◆ Igneous rocks are made when molten magma from inside the Earth cools down.
◆ Sedimentary rocks are made when broken pieces of rock, and pieces of dead plants and animals, settle in layers at the bottom of a lake or sea.
◆ Metamorphic rocks are made when igneous or sedimentary rocks are squashed or heated inside the Earth.
◆ Igneous rocks are very hard and are used for building. Sedimentary rocks can be easily dug out of quarries and so are also used for building. Metamorphic rocks often have attractive patterns and so are used for statues, monuments and important buildings. Slate is a metamorphic rock that is used for roof tiles.
◆ Most rocks are made from a mixture of minerals.
◆ Soil is mostly made up of tiny pieces of rock that have broken off from bigger rocks. It also has air, water and humus in it.
◆ Humus is made from dead and rotting plants and animals.
◆ There are different kinds of soil because there are different kinds of rock.

Remember! *(continued)*

◆ Three of the most common soils are sandy soil, clay soil and loam.
◆ Soil erosion happens when good soil is blown away by wind or washed away by water.
◆ People can cause soil erosion by cutting down trees, having too many animals on the land, or by planting the wrong crops.
◆ We can stop soil erosion by planting more trees, planting the right crops, planting different crops each year, not having too many animals, and terracing steep hillsides.

Revision Test   on Rocks, Minerals and Soil

Now that you have worked through the chapter, try this revision test. The answers are in the answer book.

1. How are igneous rocks made?
2. Name two kinds of sedimentary rock.
3. An example of a soft rock is:

   granite   marble   chalk   basalt

4. Most rocks are made from a mixture of:

   fossils   minerals   sands   soils

5. Give one common use of metamorphic rocks.
6. How is soil made?
7. How do plant roots cause weathering of rocks?
8. Why does rain water cause chemical weathering?
9. Why don't plants grow well in sandy soil?
10. Good soil can be blown away by the wind or washed away by water. What is this called?
11. Why is it bad for the soil if too many trees are cut down?
12. When we protect soil we call it soil conservation. Give two things that people can do to protect soil.

10 Water and Air

Water

The Earth is a very watery place. 70% of the surface of the Earth is covered by water. Water is very important for life on Earth. Without water there would be no plants and animals. For example, we learnt in Chapter 4 that about 70% of the human body is made up of water. We need to drink water to keep the level in our bodies at 70%. Without water we cannot survive.

Here is a drawing of the Earth showing the land coloured in green and the oceans coloured in blue. You can see that the oceans cover more of the Earth than the land covers.

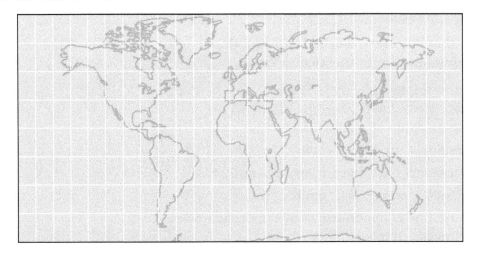

Water as a Solid, Liquid and Gas

In our daily lives we can find water in three different forms: solid, liquid and gas. Solid water is known as **ice**, liquid water is just called **water**, and when water is a gas we call it **water vapour**.

solid ice   liquid water   water vapour (gas)

Freezing, Melting and Boiling

Water **freezes** at 0 °C. When water freezes it becomes a solid and forms ice. In cold countries, this happens naturally. In hot countries, we can make ice in a refrigerator ice box or a freezer. When the temperature rises above 0 °C, the ice **melts** to form water.

If we heat water to 100 °C it will **boil** and form water vapour called **steam**. When the water vapour cools down it **condenses** to form liquid water again.

Water vapour can also form when the surface of a puddle, a lake or the sea is heated by the Sun. The water near the surface gets enough heat energy from the Sun to escape from the surface and form water vapour. This is known as **evaporation**.

When we hang wet washing on a washing line, the heat from the Sun gives the water in the clothes enough energy to evaporate. If it is a windy day the wind will carry the water vapour away before it can condense again on the washing. This is why washing dries most quickly on a sunny and windy day.

Exercise 1

a) How much of our planet is covered by water?

b) Match the three forms of water to their common names:

solid	water vapour
liquid	ice
gas	water

c) At what temperature does water boil?

d) Why does a puddle of water become smaller in the sunshine?

The Water Cycle

The total amount of water on Earth stays the same, but water is always moving between the oceans, the clouds, the land and the rivers and lakes. This is known as the **water cycle**.

Water near the surface of the sea evaporates and forms water vapour when the Sun shines on it. This water vapour cools as it rises up, so it condenses into tiny droplets, which form clouds. As more water vapour condenses, the droplets get bigger and then fall as rain. If the air is very cold it falls as snow. When water falls from the clouds as rain or snow it is called **precipitation**. The rain water that falls on land runs into rivers and then back into the ocean. And the cycle starts again.

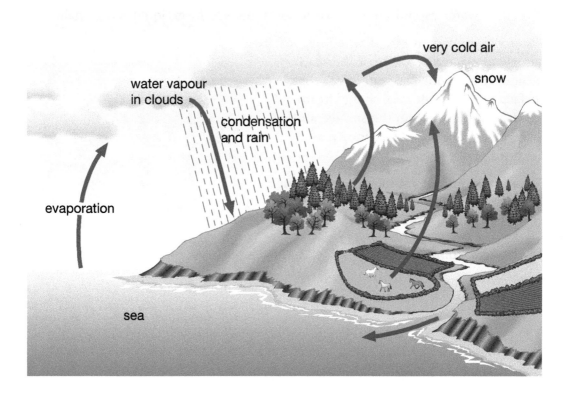

In many places around the world there is a 'rainy season' when most of the rain for the year falls. But people need water all year round, so we have to find ways of storing water so that it can be used even when there has been no rain. In many places large lakes are made by building walls called **dams** across rivers. These lakes are called **reservoirs**. In other places, water that has soaked into the ground comes back up out of the ground in a different place and starts a new stream or river. These places are called **springs**, and they usually give water all year round. In places where there are no springs people sometimes dig **wells** or **bore holes** down through the ground to where water is stored below the surface.

Salt Water and Fresh Water

Water is a **solvent**. This means that we can **dissolve** some things in it. For example, sugar dissolves in water, as we can see if we put a spoonful of sugar into a cup of water and stir it around. Chemicals called **salts** dissolve in water.

Water that has a lot of salt dissolved in it is called **salt water**. Water that has very little or no salt in it is called **fresh water**.

Only 3% of the water on Earth is fresh water. Our streams, rivers and lakes are full of fresh water, but most water on Earth is salt water. The seas and oceans are full of salt water. Humans need fresh water to live. We can't drink salt water without being ill, but animals that live in the sea have bodies that are able to live in salt water.

Water Pollution

We need fresh water to live, but sometimes our fresh water is made dirty and unsafe by **pollution**. Water pollution is when water has in it anything that is harmful to living things. There are many ways that water can become polluted. Here are some of the most common. All of them can make water too dirty to drink.

◆ People dump rubbish or garbage into rivers and the sea.
◆ People use rivers, lakes or the sea as a toilet or for emptying toilet waste.
◆ Factories and other work places dump their waste in rivers.
◆ Pesticides and other chemicals used on farms get washed into rivers.
◆ Ships dump waste and oil into water.
◆ Waste in rubbish or garbage dumps gets washed into rivers.

Dirty water can carry harmful germs that cause diseases. Water that has been polluted by people using it as a toilet, or by emptying toilet waste into it, can carry germs that cause **cholera**, **typhoid** and **dysentery**. These three diseases give people diarrhoea and vomiting, which make their bodies lose water. This can make people very ill. Two other examples of dangerous diseases carried by dirty water are **hepatitis** and **poliomyelitis**. People can be vaccinated to protect them against most diseases carried in dirty water, but the best way to stop these diseases is to make sure we keep our fresh water clean and free from pollution.

Making Water Safe for Drinking

If fresh water is not too dirty, we can make it safe for drinking by **purifying** the water. This means getting rid of the germs. Three of the most common ways of purifying water are **boiling**, **filtration** and **chlorination**.

◆ **Boiling**. Water can be boiled in a clean pot or kettle that is covered with a clean lid. Boiling the water will kill any germs that could cause disease and the water will then be clean for drinking.
◆ **Filtration**. If we strain out the solids from a liquid, we call it filtration. This can be done by pouring dirty water through a cloth into another pot or container. The cloth stops the solids in the water from passing through. We have to be careful when filtering water as germs can pass through the very small holes in cloth. When water companies or government water

departments use filtration to purify water they use special materials. For small amounts of water at home it is safer to purify the water by boiling.

◆ **Chlorination.** When there is a very large amount of water that needs to be made clean, such as a reservoir for a whole town, chemicals are added to kill the germs. The most common chemical used by water companies is chlorine. Adding chlorine to water to kill germs is called chlorination. The water that runs through our taps has usually been filtered many times to get rid of solids, and then chlorinated to kill germs.

Exercise 2

a) Here is some information about the water cycle. Fill in the gaps using these words:

    ocean   condenses   evaporates   precipitation   water vapour

    Water in the sea _____ when the Sun shines on it and forms _____ _____. This water vapour rises up and _____ into droplets, forming clouds. When there is a lot of water in the clouds it falls as rain. If the air is very cold it falls as snow. When water falls from the clouds as rain or snow it is called _____. The rain water that falls on land runs into rivers and then back into the _____, where the cycle starts again.

b) Which is more common on Earth – fresh water or salt water?
c) Give three ways that water can become polluted.
d) Give the names of two diseases that are carried by dirty water.
e) Why does boiling water help make it safe for drinking?

Air

Air is all around us. When the wind blows we can feel the movement of air, but we can't see it, smell it or taste it.

Air is a mixture of gases that have no colour or smell. The three main gases in air are **nitrogen**, **oxygen** and **carbon dioxide**. Air also has **water vapour** in it and very small amounts of other gases.

About 78% of the air around us is made up of nitrogen. Oxygen makes up about 21% of air. Carbon dioxide and other gases make up about 1% of the air around us.

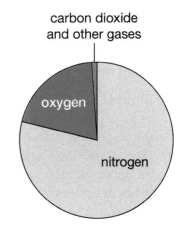

We learnt in Chapter 1 that oxygen is very important for living things. Plants and animals need oxygen for respiration. Respiration is one of the seven life processes.

Carbon dioxide is also very important for plants. Plants need carbon dioxide for photosynthesis. We learnt in Chapter 2 that plants use sunlight to turn water and carbon dioxide into sugar. Plants need the sugar for food.

The Atmosphere

The layer of air around the Earth is called the **atmosphere**. The atmosphere protects the Earth from harmful rays from space and from the Sun. The atmosphere is also like a blanket, which stops the Earth getting too hot or too cold.

The atmosphere presses down on the Earth. This pressing force is called **air pressure**. If we travel upwards through the atmosphere there is less air above us and so air pressure becomes less as we go higher. Humans can't survive if the air pressure is very low and so aeroplanes have special cabins that keep the air pressure the same as on the ground.

Air Pollution

Like water, air can also sometimes be made dirty and unsafe by pollution. Gases and particles in the air can make it harmful for humans, other animals and plants. Air pollution can come from many places. Some of the most common are:

◆ fumes from cars and trucks
◆ chemicals and smoke from factories
◆ smoke from burning rubbish or garbage
◆ smoke from burning coal
◆ smoke from cigarettes.

If people breathe in badly polluted air it can cause diseases. Common diseases caused by air pollution are **asthma** and **lung cancer**.

It is important that we keep our air as clean as possible so that it is safe for living things to breathe.

Exercise 3

a) Give the names of the three main gases in the air around us.
b) What is the name we give to the layer of air around the Earth?
c) Give three ways that air can become polluted.
d) Give the names of two diseases that are caused by polluted air.

Remember!

◆ 70% of the surface of the Earth is covered by water.
◆ Solid water is called ice, liquid water is called water, and when water is a gas we call it water vapour.
◆ Water freezes at 0 °C and boils at 100 °C.
◆ When the Sun shines on the surface of a puddle, lake or ocean it gives the water at the surface enough energy to evaporate.
◆ When water vapour cools down it condenses to form a liquid.
◆ The amount of water on Earth stays the same but it moves around in the water cycle.
◆ When water falls from the clouds as rain or snow it is called precipitation.
◆ Water is a solvent, so we can dissolve some things in it.
◆ Salt water has salts dissolved in it. The oceans are full of salt water.
◆ Humans need fresh water. Only 3% of the water on Earth is fresh water.
◆ Water pollution is when water has anything in it that is harmful to living things.
◆ Three ways we can purify water are boiling, filtration and chlorination.
◆ Air is a mixture of gases. The main gases in the air are nitrogen, oxygen and carbon dioxide.
◆ Air also has water vapour and small amounts of other gases in it.
◆ Living things need oxygen for respiration.
◆ The layer of air around the Earth is called the atmosphere.
◆ The atmosphere presses down on the Earth. This is called air pressure.
◆ Air can be made dirty and unsafe by pollution.

Revision Test   on Water and Air

Now that you have worked through the chapter, try this revision test. The answers are in the answer book.

1. How much of the Earth is covered by water?

   60%   90%   70%   55%

2. Liquid water changes to ice at 0°C. When this happens we say that the water:

   evaporates   boils   freezes   condenses   melts

3. When we boil a kettle, the water in the kettle changes from liquid to:

   solid   crystal   gas   mineral

4. What are clouds made of?

5. What is precipitation?

6. 97% of the water on Earth is salt water, 3% is fresh water. Which of these two types of water is found in the oceans?

7. When water is dirty and unsafe we say it is:

   salty   fresh   polluted   clean

8. Germs in water can be killed by adding a chemical called chlorine. This way of cleaning water is called:

   filtration   chlorination   boiling   condensing

9. Here is a pie chart showing the amounts of different gases in air.

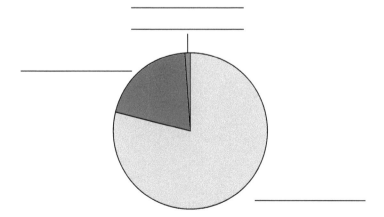

   Fill in the gaps using these three labels:

   nitrogen   oxygen   carbon dioxide and other gases

Revision Test   on Water and Air *(continued)*

10. Which gas in air is needed by all living things for respiration?

11. The layer of air that surrounds the Earth is called the:

    atmosphere   air pressure   nitrogen layer   ozone layer

12. The atmosphere presses down on the Earth. What is this force called?

11 The Environment

The Environment and Habitats

Our surroundings are called the **environment**. The environment is not the same all over the world. The Arctic is a very different place to a tropical rainforest. Humans can live in lots of different places. Humans can do this because we can wear clothes to keep us warm or cool, and we can build houses to protect us from the weather. We can also eat lots of different types of food. Plants and animals can't live anywhere. For example, polar bears can't live in a hot desert and a parrot would not survive in the Arctic. Where a plant or animal lives is called its **habitat**.

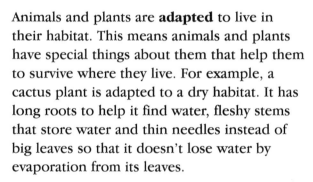

thin needles

fleshy stems

Animals and plants are **adapted** to live in their habitat. This means animals and plants have special things about them that help them to survive where they live. For example, a cactus plant is adapted to a dry habitat. It has long roots to help it find water, fleshy stems that store water and thin needles instead of big leaves so that it doesn't lose water by evaporation from its leaves.

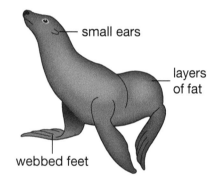

small ears

layers of fat

webbed feet

Sea lions that live near to the Antartic are adapted to live in a cold habitat. They have layers of fat to keep the body warm, small ears to cut heat loss, and a streamlined body and webbed feet to help with swimming.

A camel is adapted to live in a dry, sandy habitat. It has broad flat feet and long legs so that it doesn't sink into the sand. Camels have long eye lashes and nostrils that can close so that they don't get sand in their eyes or nose. They also have a hump which stores fat. Camels store this fat for food which means they can go for a long time without eating. They can also go for a long time without drinking because they can drink a lot of water in one go and store it in their bodies.

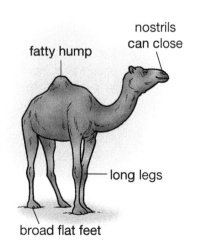

nostrils can close

fatty hump

long legs

broad flat feet

Exercise 1

a) What is the name of the place where a plant or an animal lives?
b) Give two reasons why humans can live in lots of different places.
c) Give one adaptation that a cactus has for living in a hot and dry habitat.
d) Give one adaptation that a seal has for living in a cold habitat.
e) Can you think of an adaptation that a fish has for living in the sea?

Looking After the Environment

We learnt in Chapter 10 that humans can pollute water and air, which makes it harmful to humans and other living things. Water and air are two of the most important parts of our environment. If we pollute the environment too much we will kill plants and animals, and destroy their habitats.

Acid Rain

One way pollution damages the environment is by causing **acid rain**. We learnt in Chapter 9 that rain water is naturally slightly acidic. But rain water is much more acidic if the air is polluted with chemicals that have sulphur and nitrogen in them. Coal-burning power stations, petrol engines and diesel engines all make pollution that has sulphur and nitrogen chemicals in it. The sulphur and nitrogen chemicals dissolve in rain water to make it more acidic. This is called acid rain. Acid rain can kill plants and pollute soil.

Greenhouse Gases

The atmosphere acts as a blanket around the Earth. It lets in just the right amount of warmth from the Sun. It also keeps the Earth cool by letting some of the heat that is reflected off the Earth escape back into space. Carbon dioxide and other gases in the atmosphere stop some of the heat from escaping. Humans are increasing the amount of carbon dioxide in the atmosphere. Coal-burning powers stations give out a lot of carbon dioxide as well as other gases. Cars, trucks and aeroplanes all give out carbon dioxide. Carbon dioxide and other gases are building up in the atmosphere, trapping more heat. These gases are called **greenhouse gases**.

Because the amounts of greenhouse gases are increasing, the Earth's atmosphere is getting warmer. It is getting warm enough to make habitats change in some places, which means some plants and animals can't survive.

Also, as the atmosphere gets warmer it changes the weather so that some places get more rain and some places get less rain than they are used to. This can cause floods in some places and **drought** in other places. A drought is when there is no rain for a long period of time.

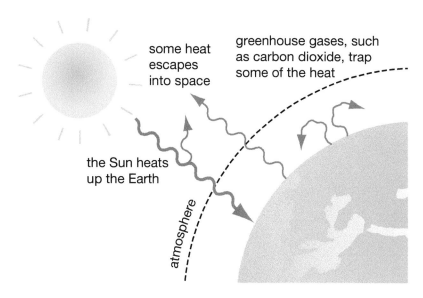

some heat escapes into space

greenhouse gases, such as carbon dioxide, trap some of the heat

the Sun heats up the Earth

atmosphere

If we make too much acid rain, or if we keep increasing greenhouse gases, or if we pollute the environment in other ways, we will destroy habitats and kill plants and animals. If we destroy the habitat for one living thing it will harm other living things too, because living things are linked together through **food chains**.

Exercise 2

a) Which two chemcials dissolve in rain water to make acid rain?
b) Why is acid rain bad for the environment?
c) What is the name we give to gases that make the Earth's atmosphere warmer?
d) How do gases like carbon dioxide make the atmosphere warmer?

Food Chains

Lots of different plants and animals can live together in the same habitat. They all need food for growth and energy. Plants use sunlight to make their food from water and air. Animals eat plants and other animals to get their food. Living things are all linked together in food chains. The diagram on the next page shows a food chain.

In this food chain the plant is called the **producer**, because it produces (makes) its own food. The rabbit is called the **consumer** because it consumes (eats) another living thing. When the rabbit dies it rots back into the soil and becomes humus, as we learnt in Chapter 9. Micro-organisms like bacteria and fungi help the rabbit to become humus. Another word for rot is 'decompose', so the micro-organisms are called **decomposers**.

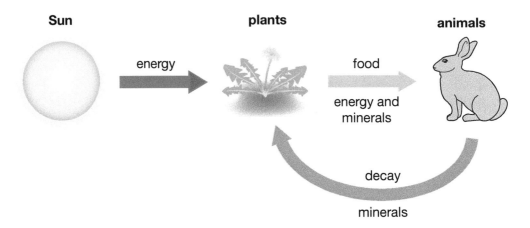

Here is another, longer food chain.

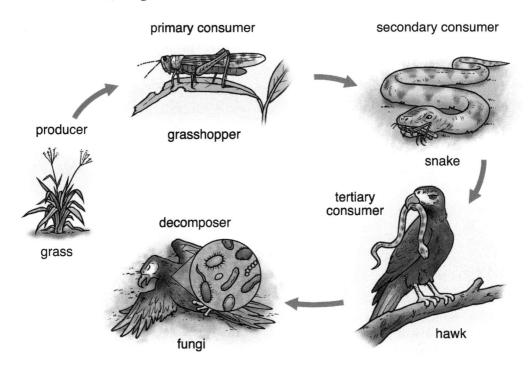

The grasshopper eats the grass, the snake eats the grasshopper, and the hawk eats the snake. When the hawk dies it decomposes into the soil. In this food chain there are three consumers, so they are called the **primary consumer**, the **secondary consumer** and the **tertiary consumer**. Normally we write food chains in a line like this:

grass → grasshopper → snake → hawk → fungi

We can now see that living things are linked through food chains, so if we damage the habitat for one living thing it will affect other living things too. For example, if factories and coal-burning power stations make too much acid rain, it will kill grass. If there is less grass there will be fewer grasshoppers because their habitat will have been destroyed and they won't have enough food. This means less food for snakes so there will be fewer snakes. In turn, this means less food for hawks, so they will also start to die out.

Endangered Species

Some living things have almost disappeared from the Earth because their habitats have been destroyed. For example, the orang-utan is dying out in Malaysia and Indonesia because its habitat is being destroyed. Large parts of the rainforest where orang-utans live have been cleared to grow palm oil crops. Other living things have almost disappeared because humans have killed too many of them for food. For example, some types of whale have nearly died out because too many of them have been caught by humans. When a living thing becomes very rare it is called an **endangered species**.

Exercise 3

Look at this food chain.

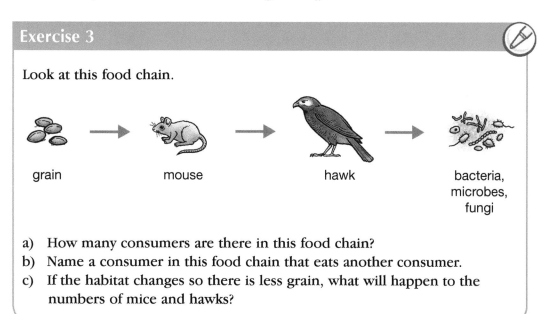

grain   mouse   hawk   bacteria, microbes, fungi

a) How many consumers are there in this food chain?
b) Name a consumer in this food chain that eats another consumer.
c) If the habitat changes so there is less grain, what will happen to the numbers of mice and hawks?

Remember!

◆ Our surroundings are called the environment.

◆ Where a plant or animal lives is called its habitat.

◆ Animals and plants are adapted to live in their habitat.

◆ If we pollute the environment too much we will kill plants and animals and destroy their habitats.

◆ Sulphur and nitrogen from pollution in the air dissolve in rain water to make acid rain. This can kill plants and pollute soil.

◆ Carbon dioxide from pollution traps heat in the atmosphere and makes the Earth's atmosphere warmer. Carbon dioxide and other gases that do this are called greenhouse gases.

◆ The amounts of greenhouse gases are increasing and the world is becoming warmer. This is changing some habitats so that some plants and animals can't survive.

◆ All living things are linked through food chains.

◆ Most food chains start with a plant. The plant is called the producer because it makes its own food.

◆ Animals in a food chain are called consumers because they eat other living things.

◆ When a living thing dies it is turned into humus in the soil by micro-organisms called decomposers.

◆ If we kill a plant or animal in a food chain, or destroy its habitat so that it cannot survive, other animals in the food chain will have less food and they could die out too.

◆ When a living thing becomes very rare it is called an endangered species.

Revision Test on The Environment

Now that you have worked through the chapter, try this revision test. The answers are in the answer book.

1. Name two adaptations that a camel has for living in a hot, dry desert.
2. Name one adaptation that a polar bear has for living in a cold habitat.
3. Give one adaptation that helps owls hunt at night.
4. What are greenhouse gases?
5. Why do greenhouse gases make the atmosphere warmer?
6. Why are greenhouse gases and acid rain bad for the environment?
7. Look at the food chain below.

   a) Name the producer in this food chain.
   b) How many consumers are there?
   c) Name a consumer that eats another consumer.
   d) What would happen to the number of snakes if the habitat changed and there were no leaves?

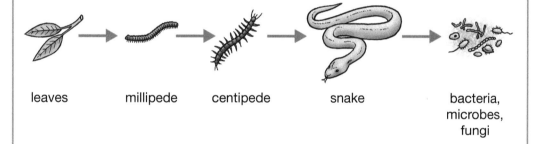

leaves   millipede   centipede   snake   bacteria,
                                                                microbes,
                                                                fungi

8. Look at the food chain below.

   What would happen to the number of krill in this food chain if too many whales were killed by humans?

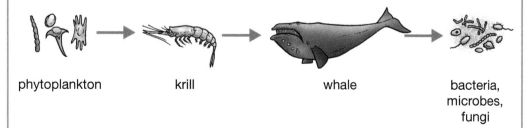

phytoplankton   krill   whale   bacteria,
                                                           microbes,
                                                           fungi

9. How could destroying grassland kill owls?
10. What is an endangered species?

12 The Earth in Space

The Solar System

The Earth is a sphere that moves around the Sun. You will have learnt in your Maths lessons that a sphere is the shape of a round ball. People used to think that the Earth was still and that the Sun moved around the Earth, but we now know that the Earth moves around the Sun in a path called an orbit. The shape of the orbit is a slightly stretched-out circle, called an **ellipse**.

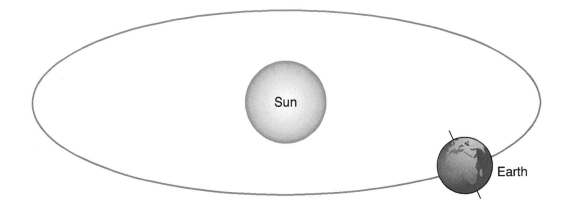

The Sun is a star. Stars are big balls of hot gas that give out heat and light. The Earth is a planet. Planets orbit around a star and they don't give out light. We can see planets because they reflect light from the star they orbit around.

The word 'solar' means 'to do with the Sun'. The **Solar System** is the name given to the Sun and all the things that move around it. The Earth is not the only planet moving around the Sun, there are other planets in the Solar System too. The diagram opposite shows their names.

The table gives some facts about the planets in the Solar System.

From the data in the table you can see that Earth has a diameter of 12 682 km and is 150 million km from the Sun. The table also tells us that it takes 365.25 days for the Earth to orbit the Sun. Each orbit of the Sun is one year, so for the Earth one year is 365.25 days.

You can also see from the data in the table that Pluto is smaller than the other planets. It has a diameter of only 2300 km. Many scientists think that Pluto is too small to be a proper planet and so they don't count it as a planet.

not drawn to scale

Name of planet	Diameter of the planet (km)	Average distance from the Sun (million km)	Time taken for one orbit around the Sun (number of Earth days)
Mercury	4 880	58	88
Venus	12 100	108	225
Earth	12 682	150	365.25
Mars	6 720	228	687
Jupiter	141 920	778	4 344
Saturn	120 000	1427	10 755
Uranus	51 160	2871	30 664
Neptune	49 860	4497	60 148
Pluto	2 300	5914	90 717

Exercise 1

Look at the data in the table on page 105, and then answer these questions.

a) Which is the largest planet in the Solar System?
b) What is the diameter of the largest planet?
c) What is the diameter of Earth?
d) How far is Earth from the Sun?
e) How far is Earth from Mars?
f) How far is Mercury from the Sun?
g) How many Earth days does it take for Mercury to orbit the Sun?

Day and Night

As the Earth moves around the Sun, it spins on its axis. The axis is the line through the centre of the Earth from the North Pole to the South Pole.

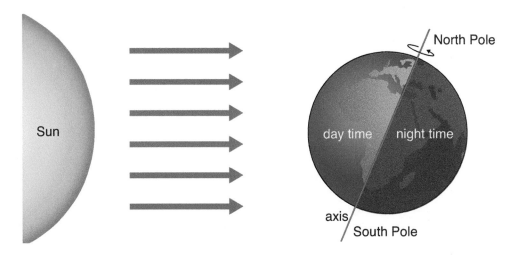

The Earth takes 24 hours to spin round once, so 24 hours is the length of one day. As the Earth spins, the side of the Earth facing the Sun is lit up and so it is daytime on that side of the Earth. The side of the Earth facing away from the Sun is in darkness, so it is night time on that side. The Earth's axis is tilted to one side.

The Four Seasons – Spring, Summer, Autumn, Winter

There is an imaginary line running around the middle of the Earth called the **equator**. The half of the Earth above the equator is called the northern hemisphere, and the half below the equator is called the southern hemisphere. If you live close to the equator the temperature is about the same all year, but if you live far from the equator in the northern or southern hemisphere, there is a big difference in temperature between summer and winter. In the summer it is much warmer than in the winter.

Because the Earth's axis is tilted, there is a time of year when the northern hemisphere is pointing towards the Sun and the southern hemisphere is pointing away from the Sun. When this happens it is summer in the northern hemisphere because the northern hemisphere gets more heat and light from the Sun. It is winter in the southern hemisphere because it is pointing away from the Sun and so gets less heat and light.

As the Earth orbits around the Sun the positions change. When the Earth gets to the other side of the Sun, the southern hemisphere points towards the Sun and the northern hemisphere points away. Now it is summer in the southern hemisphere and winter in the northern hemisphere.

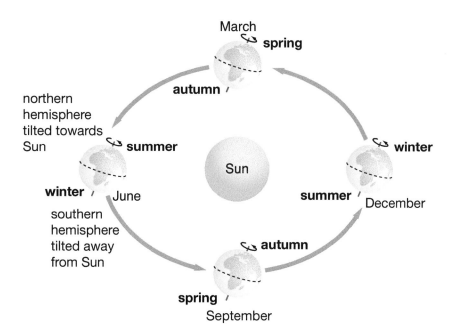

The Earth and the Moon

We learnt earlier in this chapter that the Earth orbits around the Sun. The Moon orbits around the Earth in the same way. It takes about 28 days for the Moon to orbit the Earth.

If you look at the Moon on different nights it seems to change shape. This is because we can't always see all of the Moon. We can only see the parts of the Moon that reflect light from the Sun towards the Earth. So as the Moon orbits the Earth, different parts of it can be seen from the Earth.

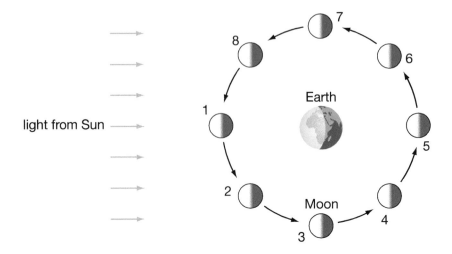

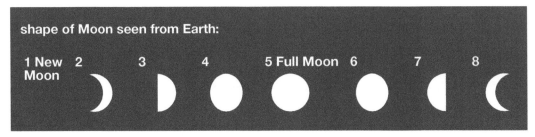

The different shapes we can see from Earth are called the **phases** of the Moon.

Why Do We Always See the Same Side of the Moon?

As the Moon moves around the Earth, it also spins on its axis. The Moon spins round once on its axis every time it orbits the Earth. In other words, it takes the Moon exactly the same time to spin round once on its axis as it takes the Moon to orbit the Earth. This means the same side of the Moon always faces the Earth.

Exercise 2

a) How long does it take the Earth to spin around once on its axis?
b) When it is day time on one side of the Earth, why is it night time on the other side?
c) When the northern hemisphere is tilted towards the Sun, is it summer or winter in the northern hemisphere?
d) About how long does it take the Moon to orbit the Earth?
e) Why does the Moon look different on different nights as it orbits the Earth?
f) Why do we always see the same side of the Moon?

Beyond Our Solar System

Stars group together to form a galaxy. Our Sun is one star in a very large group of stars that make up the Milky Way galaxy. The Milky Way galaxy is huge. There are about 200 billion other stars like our Sun in the galaxy, and the nearest one to us is so far away that it takes light just over four years to travel from the star to the Earth, even though light travels at 300 000 kilometres per second.

The Milky Way galaxy is shaped like a spiral. Our Sun is positioned in one arm of the spiral. These drawings show the position of the Sun in the Milky Way galaxy. Remember that in the real galaxy there are billions of stars.

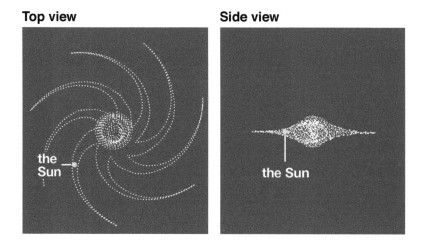

The Milky Way galaxy is so big that it takes light 100 000 years to cross from one side to the other.

Beyond Our Galaxy

As telescopes have got bigger and better, scientists have been able to see farther into space. Scientists now estimate that there are billions of other galaxies like the Milky Way galaxy, and each one has billions of stars. The Universe is made up of all the galaxies and the empty space between them.

Unless we have a telescope we can't see most of the stars from Earth because they are too far away. But on a clear night from anywhere on the Earth we can see up to 2500 individual stars with our own eyes. Next time there is a clear night where you live, look at a star and remember that, because the star is very far away, you are seeing light that left the star many years ago.

Exercise 3

a) What is a galaxy?
b) What is the name of the galaxy we are part of?
c) What is the shape of our galaxy?
d) Up to how many individual stars can we see on a clear night without a telescope?

Remember!

◆ The Earth is a sphere that moves around the Sun in a path called an orbit. The shape of the orbit is a slightly stretched-out circle, called an ellipse.
◆ The Sun is a star and the Earth is a planet.
◆ The Solar System is the name given to the Sun and all the things that orbit around it.
◆ It takes the Earth 365.25 days to orbit the Sun, so 365.25 days is one year.
◆ The Earth spins once on its axis every 24 hours, so 24 hours is one day.
◆ The imaginary line around the middle of the Earth is called the equator.
◆ The half of the Earth above the equator is called the northern hemisphere, and the half below the equator is called the southern hemisphere.
◆ The Earth's axis is tilted, so when one hemisphere points towards the Sun the other hemisphere points away from the Sun.
◆ It is summer in the hemisphere pointing towards the Sun and winter in the hemisphere pointing away from it.
◆ It takes the Moon about 28 days to orbit the Earth.
◆ It takes the Moon exactly the same time to spin once on its axis as it takes for it to orbit the Earth. This means the same side of the Moon always faces the Earth.
◆ Our Sun is one star in the Milky Way galaxy. There are billions of stars in the galaxy.
◆ Scientists estimate that there are billions of galaxies in the Universe.

Revision Test on The Earth in Space

Now that you have worked through the chapter, try this revision test. The answers are in the answer book.

1. What shape are the Sun, the Earth, and the other planets?
2. What does the Earth orbit around, and how long does it take to make one orbit?
3. Put these planets in order of size, with the smallest first:

   Earth   Mars   Jupiter

4. It takes the Earth 24 hours to spin once on its axis. What do we call this length of time?
5. When the northern hemisphere is tilted towards the Sun, is it summer or winter in the southern hemisphere?
6. When it is summer at the South Pole, why is it winter at the North Pole?
7. What does the Moon orbit around?
8. When no part of the Moon reflects light towards Earth, how much of the Moon can we see?
9. What is the Milky Way?
10. How long does it take light to travel to Earth from the nearest star in our galaxy?

Revision Tests

Now that you have worked your way through all the chapters in this book, try these revision tests. The answers are in the answer book.

Revision Test 1

1. Name the seven life processes.
2. Which life process makes new living things?
3. Which part of a plant makes food for the plant?
4. Name three ways that seeds and fruits can be dispersed.
5. Name the three types of blood vessel in a human body.
6. Which human organ produces bile to help break down food?
7. What test could we do to see if a food has starch in it?
8. Name the five things that forces can make objects do.
9. Name a force that slows down a moving object.
10. Will these two magnets attract each other or repel each other?

11. What units do we use to measure force?
12. Name the seven colours in the spectrum of white light.
13. Here is a drawing of the human eye. What is the part labelled X called?

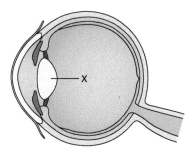

14. Which travels faster – light or sound?
15. What is the name of the electrical component labelled X in this diagram?

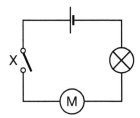

16. Name the three main types of rock.
17. Why don't plants grow well in sandy soil?
18. Why does cutting down trees cause soil erosion?
19. What is solid water called?
20. Here is a drawing of the water cycle. Which letter shows where most evaporation is happening?

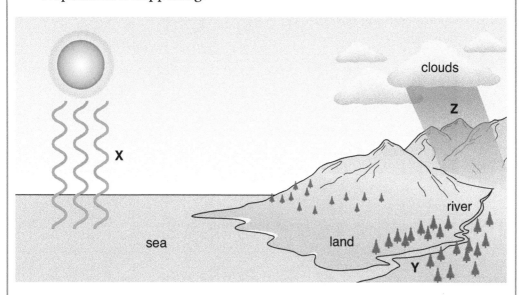

21. What is the gas in air that living things need for respiration?
22. Give the name of a greenhouse gas.
23. Look at this food chain.
    Give the name of the producer in this food chain.

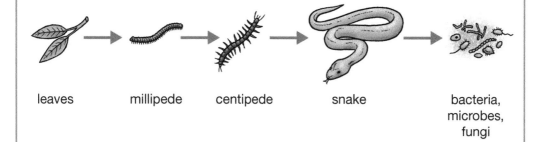

leaves   millipede   centipede   snake   bacteria, microbes, fungi

24. When the southern hemisphere is tilted towards the Sun, is it summer or winter in the northern hemisphere?
25. How long does it take for the Earth to spin once on its axis?

Revision Test 1 *(continued)*

Multiple Choice Questions

26. Which of these four animals is not a mammal?

    A   cat

    B   snake

    C   cow

    D   mouse

27. In this drawing, the part of the plant labelled X is:

    A   the ovary

    B   the pistil

    C   the stamen

    D   the petal

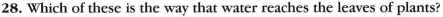

28. Which of these is the way that water reaches the leaves of plants?

    A   stem → roots → leaves

    B   roots → stem → leaves

    C   leaves → stem → roots

    D   roots → leaves → stem

29. Which of these is not an organ of the human body?

    A   lung

    B   heart

    C   finger

    D   liver

30. Where does most digestion take place in the human body?

    A   small intestine

    B   stomach

    C   large intestine

    D   oesophagus

31. Which of these nutrients gives us lots of energy?

    A   proteins

    B   minerals

    C   vitamins

    D   carbohydrates

32. Which of these materials is a good electrical insulator?

    A   copper

    B   aluminium

    C   plastic

    D   steel

33. How much of the human body is made up of water?

    A   about 15%

    B   about 50%

    C   about 40%

    D   about 70%

34. Which one of these kinds of energy is stored in a stretched elastic band?

    A   chemical energy

    B   strain energy

    C   kinetic energy

    D   gravitational potential energy

35. When heat spreads through a gas from hot to cold it is called:

    A   convection

    B   radiation

    C   insulation

    D   conduction

36. Which of these statements is true?

    A   Sedimentary rocks are made when lava from a volcano cools.

    B   Sedimentary rocks are made when other rocks are heated inside the Earth.

    C   Sedimentary rocks are made when bits of rock and dead plants and animals form layers.

    D   Sedimentary rocks are made when other rocks are weathered by the wind and rain.

37. At what temperature does water freeze?

    A   50 °C

    B   −100 °C

    C   0 °C

    D   100 °C

38. Which gas makes up about 78% of the air around us?

    A   oxygen

    B   carbon dioxide

    C   hydrogen

    D   nitrogen

39. The Sun and all the things that orbit around it together are known as:

    A   the Universe

    B   the Galaxy

    C   the Milky Way

    D   the Solar System

40. Which of these is the name given to the imaginary line around the middle of the Earth?

    A   the equator   B   the hemisphere

    C   the orbit   D   the Arctic circle

Revision Test 2

1. Living things use food to stay alive. What is this life process called?
2. To which kingdom of living things does a snake belong?
3. Look at these drawings of plant roots.

   Which one shows storage roots?

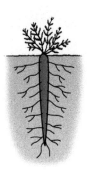  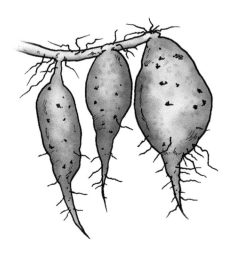

4. Name two ways that a plant can be pollinated.
5. Look at this drawing of a leaf.

   What is the name of the part labelled X?

6. Which type of blood vessel carries blood away from the heart?
7. Name a food that gives us protein.
8. Name a disease that is spread in dirty water.
9. Give the name of a non-contact force.
10. John has a mass of 50 kg. If the force of gravity on one kilogram is 10 N, what is the force of gravity on John?

11. Here is a drawing of the particles in three substances. Which substance is a gas?

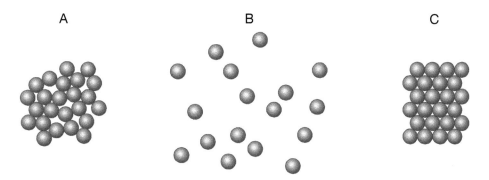

A   B   C

12. Give an example of a luminous object.

13. When there is a thunder storm, the lightning and the thunder happen at the same time. Why do we see the lightning before we hear the thunder?

14. In this drawing, the girl can see the table because the light is reflected from it into the eye. Draw a light ray showing how the light travels from the light source to the eye.

15. Draw the electrical circuit symbols for a bulb and a battery.

16. How are igneous rocks made?

17. Give two ways that water can be polluted.

18. Name three ways of purifying water.

19. About how much of the water on Earth is fresh water?

20. What is the atmosphere?

21. How do greenhouse gases make the Earth warmer?

22. Name two adaptations that a cactus plant has for living in a hot and dry habitat.

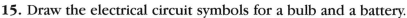

Revision Test 2 *(continued)*

23. How many days does it take the Earth to make one complete orbit of the Sun?
24. Why do we always see the same side of the Moon?
25. What is the name of our galaxy?

Multiple Choice Questions

26. Which of these is a vertebrate?
    - A   an animal with a backbone
    - B   an animal without a backbone
    - C   a flowering plant
    - D   a bacterium
27. Which of these is the sticky part of the flower that collects pollen?
    - A   ovary
    - B   anther
    - C   stigma
    - D   petal
28. Which gas does blood give to the air in the lungs to be breathed out?
    - A   oxygen
    - B   carbon dioxide
    - C   nitrogen
    - D   hydrogen
29. Which of these foods is a good source of carbohydrate?
    - A   beans
    - B   carrots
    - C   bread
    - D   tomatoes
30. In this tug-of-war game, no one is moving.

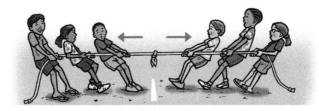

How would you describe the forces acting on the rope?
    - A   unbalanced forces
    - B   pushing forces
    - C   balanced forces
    - D   turning forces

Revision Test 2   *(continued)*

31. A force can make an object turn around a point. What name is given to the point an object turns around?

   A   an axis

   B   an angle

   C   a lever

   D   a pivot

32. Which of these is the energy of movement?

   A   chemical energy

   B   kinetic energy

   C   gravitational potential energy

   D   electrical energy

33. Which of these materials is a good conductor of heat?

   A   plastic

   B   wood

   C   metal

   D   rubber

34. Heat from the Sun gets to the Earth by:

   A   convection

   B   radiation

   C   conduction

   D   reflection

35. Which of these materials is a good electrical insulator?

   A   rubber

   B   aluminium

   C   iron

   D   copper

36. A farmer is looking at the soil. Which soil will be best for growing crops?

   A   sandy

   B   clay

   C   loam

   D   gravel

37. When liquid water changes to gas at 100 °C, we say that the water:

   A   freezes

   B   condenses

   C   melts

   D   boils

$\rightarrow$

38. Here is a drawing of the Earth orbiting the Sun.

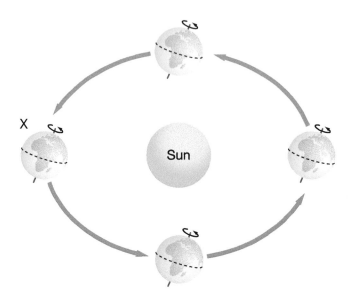

When the Earth is at point X, it is:

A   winter in the northern hemisphere and summer in the southern hemisphere

B   summer all over the Earth

C   summer in the northern hemisphere and winter in the southern hemisphere

D   autumn in the northern hemisphere and spring in the southern hemisphere

39. Which is the largest planet in the Solar System?

A   Mercury

B   Jupiter

C   Earth

D   Mars

40. Which of these is a galaxy?

A   the Sun and all the things that move around it

B   the orbit of the Moon around the Earth

C   billions of stars grouped together

D   the distance from the nearest star to Earth

1. Which life process lets living things use oxygen to get energy from food?
2. All living things get bigger. What is this life process called?
3. To which kingdom of living things does an earthworm belong?
4. Plant roots do three main jobs. What are they?
5. How long does it normally take for a human baby to be born after the mother's egg has been fertilised?
6. Name the five main groups of nutrients.
7. Diseases that pass from one person to another are called communicable diseases. Give two ways that a communicable disease can be spread.
8. When an apple falls from a tree, what force makes it fall to the ground?
9. Rosie has two magnets. What will happen to the magnets if Rosie puts the two south poles together?
10. Why is oil a non-renewable source of energy?
11. Why is solar power a renewable source of energy?
12. What is the name of the part of the ear labelled X in this drawing?

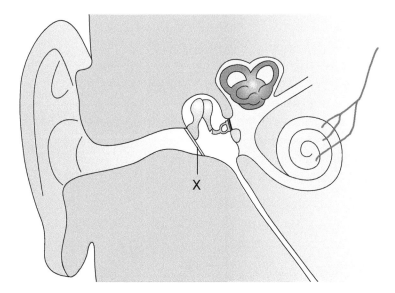

13. When light rays pass from air to water they change direction slightly. What name do we use to describe light rays changing direction like this?
14. When sound is reflected, what name do we give to the sound we hear?
15. What units are used to measure loudness?

16. Here are four circuit diagrams. Each circuit has a bulb in it. In which of these circuits will the bulb light up?

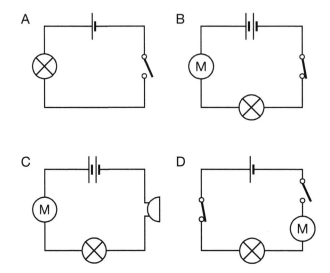

17. How are metamorphic rocks made?
18. Give two ways that people can cause soil erosion.
19. Name two types of physical weathering of rocks.
20. Name three important gases in air.
21. The layer of air around the Earth presses down on the Earth. What is the name of this pressing force?
22. Give two adaptations that a polar bear has for living in the Arctic.
23. Look at this food chain.

    Name two consumers in this food chain.

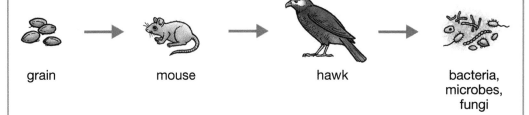

24. What does the Moon orbit around?
25. How many days does it take for the Earth to spin once on its axis?

(continued)

Multiple Choice Questions

26. Which of these is *not* one of the seven life processes?
    A   movement
    B   hearing
    C   sensitivity
    D   nutrition

27. Which of these is the part of the flower that makes pollen?
    A   stalk
    B   sepals
    C   anther
    D   ovary

28. When plant seeds start to grow it is called:
    A   germination
    B   reproduction
    C   dispersal
    D   pollination

29. Which of these is *not* a job done by your skeleton?
    A   It protects your body parts.
    B   It helps you digest food.
    C   It supports your body.
    D   It lets you move.

30. Which of these is a test to see if a food has fat in it?
    A   the iodine test
    B   Benedict's test
    C   the grease spot test
    D   the biuret test

31. Which one of these things will be attracted to a magnet?
    A   a wooden pencil
    B   a can made of aluminium
    C   a plastic spoon
    D   a steel nail

32. Which one of these instruments can be used to measure force?
    A   a newtonmeter
    B   a voltmeter
    C   a thermometer
    D   an ammeter

33. Which of these types of energy is made by a vibrating object?
    A   chemical energy
    B   electrical energy
    C   sound energy
    D   light energy

34. When heat from a fire travels along a metal poker it is called:
A   convection
B   conduction
C   radiation
D   insulation

35. What name do we give to a material that doesn't allow light to pass through it?
A   transparent
B   opaque
C   luminous
D   translucent

36. Which of these materials is a good electrical conductor?
A   iron
B   plastic
C   wood
D   rubber

37. When water changes from a gas to a liquid we say the water:
A   freezes
B   evaporates
C   condenses
D   melts

38. At what temperature does water boil?
A   0°C
B   100°C
C   −100°C
D   20°C

39. Which of these causes acid rain?
A   gases from pollution dissolved in rain water
B   rubbish and garbage thrown into rivers
C   physical weathering of rocks
D   oil pollution in the sea evaporating to make rain

40. What is the biggest object in the Solar System?
A   Earth
B   Jupiter
C   the Sun
D   Saturn